JOURNEY
to
GIZA

By

DHAN

Synopsis

The narrator, having helped construct the great pyramid at Giza, sets out to reveal to the reader just how it was done.

He succeeds.

Along the way he embroils the reader, along with friends and others, in the troubles of his time.

While, all the while, searching for an answer to a simple question. How did being become?

Thereby beginning a journey into existentialism.

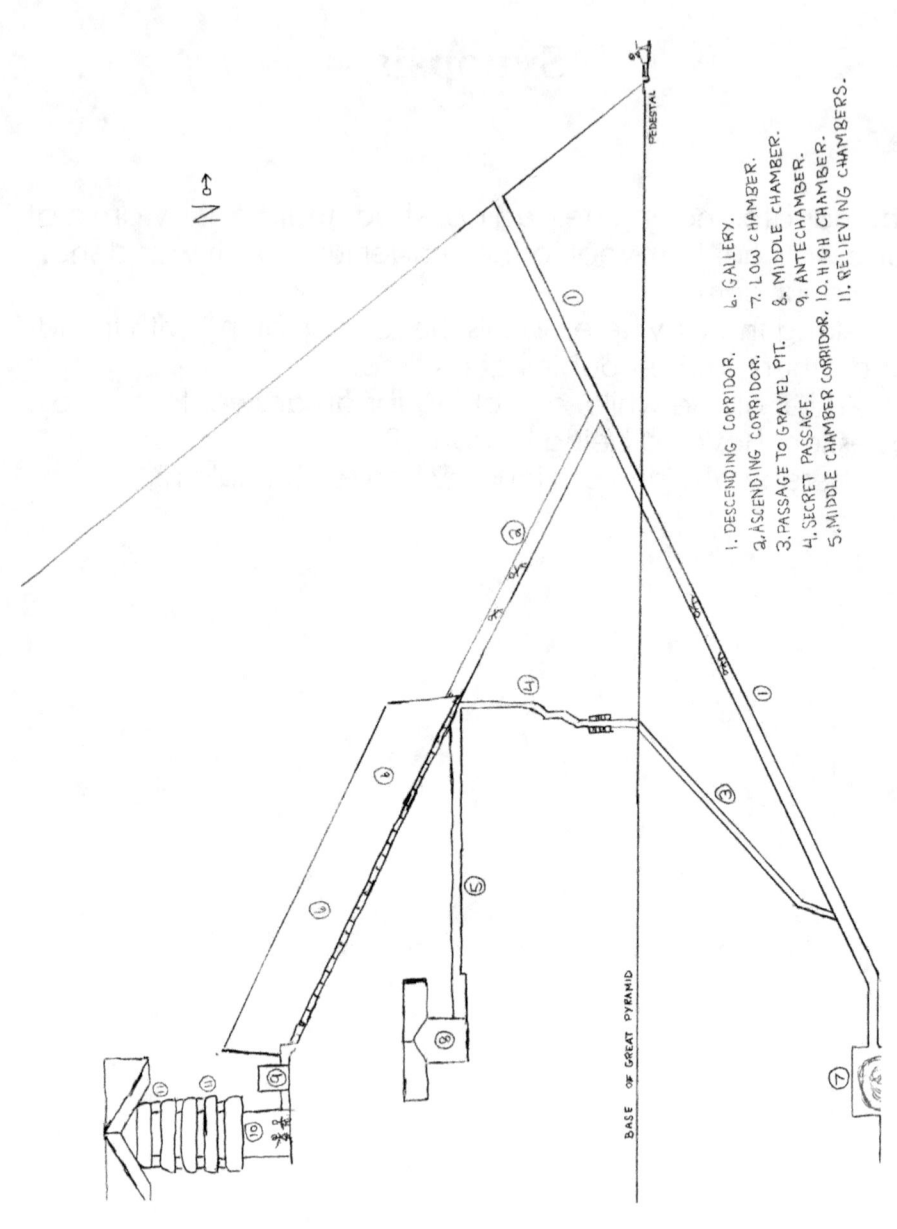

N ↑

PEDESTAL

1. DESCENDING CORRIDOR.
2. ASCENDING CORRIDOR.
3. PASSAGE TO GRAVEL PIT.
4. SECRET PASSAGE.
5. MIDDLE CHAMBER CORRIDOR.
6. GALLERY.
7. LOW CHAMBER.
8. MIDDLE CHAMBER.
9. ANTECHAMBER.
10. HIGH CHAMBER.
11. RELIEVING CHAMBERS.

BASE OF GREAT PYRAMID

The First Scroll

You began a journey when you unrolled this scroll.

It will be a journey to Giza and into the greatest of pyramids.

It will be trip back to a time when I was a young man. A young man who pulled stone on that structure.

Many a night I have sat by lamplight constructing our way, word by word; as men construct a road, stone by stone.

I choose to write because I am a scribe, it is what I do. I am also a white haired widower with time on my hands.

I choose to write because of foolish tales I have heard. There are foreigners saying that this great work was raised up by abused slaves.

I will tell you how it was really created.

I will tell you how it was created in keeping with our history of pyramid building.

I will tell you how it was created with an economy of words, just as that more solid one was created with an economy of stone.

Many have seen those stones which attract other stones of their kind, or push them away.

Many have manipulated such stones themselves. So all know, this is no jugglers' trick.

There is power in stone.

In our wood poor land many know of this power in stone. There are those whose families have, for generations, shaped and built with stone.

In our nearly treeless land stone and brick are the main building materials. Only the rich can afford to live in houses of wood.

We begin with its beginnings.

A limestone quarry was found. Limestone is easy to quarry and easy to shape. This great pyramid will be built of limestone.

The quarry was close to the great river, as it should be. Ships will bring such basics as grain and rope to those who will build this pyramid.

In both of these our land is rich. Both of these we export to other lands around the great green sea.

The ships will bring other things as well.

Pure white limestone is the final coat this pyramid will wear. This white stone will come, the pieces already shaped, from another quarry.

Granite will be needed for one tomb to be erected within this structure. Its power to support great weight and span great distance will be wanted. It will come, the pieces already shaped, from a third quarry.

The power in soapstone will help build this structure. Heavy loads will glide over its waxy surface while the ropemens' feet grip rougher stone.

A great many ropemen will be needed.

Men of bone and muscle yet men who know a pyramid is so shaped that its bulk leans into its own center, because inner stone is so angled.

Time, and the tremors of an aging world, will only serve to make that vast bulk more compact.

These men take pride in building for eternity.

The way lesser men take pride in name.

It has been a great many years since the last pyramid was constructed. There are many who wonder if there are enough men, yet living, who possess the skills needed to build another.

Let me try to show what concerns them.

Assume that day has come when the capstone is set in place. The pyramid is not yet finished.

Stonemasons gather on the topmost tier.

The time for talking must end. On all four sides the stonemasons begin cutting into the white limestone at the base of the capstone.

As work progresses they stop, now and then, to take measurements.

In time it becomes evident what these stonemasons are doing.

They are cutting a flight of smaller steps into the many tiers of the structure.

They are joined by stonecutters. The best of these will, one day, be apprentice stonemasons.

The stonecutters view the steps as triangular rows of stone. Starting with the topmost row, they cut away these triangles of stone at their base.

Some men remove rubble.

Others, using sandstone blocks, smooth the surface.

All wonder if these steps will descend as true as steps once carved by their fathers and their grandfathers.

Three tombs were planned.

Only one will be completed.

No one knows which one.

The first, the low chamber, will be carved out of the bedrock deep beneath the structure.

It will be reached by way of a paneled corridor that descends to it from ground level.

Each panel will be made of smooth granite.

At the low end of this descending corridor, in its west wall, will be an entrance into a steep passageway.

This passageway will connect the descending corridor to a gravel pit. A gravel pit that will be buried beneath the structure.

At the upper end of this steep passageway will be two other small, mystical, passages.

Pharaoh will not use either one of these mystical passages until after his death.

Ships will be entombed in pits, close to the finished pyramid.

Some believe one mystical passage allows pharaohs' soul access to another world. There another great river and ships, like some entombed, await pharaoh.

They believe the other mystical passage allows pharaohs' soul access to this world.

They say they sense, at times, the presence of ghostly craft on our own great river.

They claim they hear, now and then, the haunting cry of a lookout from a bow.

If the low chamber is not completed, the mystical passages will remain closed. The entrance into the steep passageway will be concealed with a granite wallpanel.

There is something else to tell. It belongs at the beginning.

There is a ridge within the pyramids perimeter.

It will be buried within the pyramid, after its shape is altered and it is given new form.

There is power in shapes and forms.

Sand, carried by the prevailing north wind, has eroded the sides of the ridge. They are concave.

The ridge is sloped.

Its south end is its high end, as with our great river.

Stone was removed from the south face until it was at a right angle to the east face.

The same was done to the north face.

The east face points due north.

An ascending corridor will be constructed. Its entrance will be in the ceiling of the descending one. It will ascend to the low end of the ridge. It will give access to structures built in, and on, the ridge.

The upper section of the ascending corridor will be constructed of stones running parallel to its line of ascent.

The lower section of the ascending corridor will be cut through large limestone blocks. Blocks that will stop and hold a plugstone between their solid walls. Walls too close together for it to pass.

The slope of the ridge has been cut into a long staircase.

The staircase is divided by an aisle.

At the top of the aisle stands a large block of stone, fashioned from the stone of the ridge.

One day this aisle will be enclosed in a long, inclined gallery. The block of stone will be a great step at the gallerys' high end.

The gallerys' aisle will seem a continuation of the ascending corridor.

The inclined gallery will hold, then launch, the pyramids' plugstones. To launch the plugstones it must be in line with the ascending corridor.

This will be something unique to this structure. Plugstones will be launched from within a pyramid, in order to block entry into that same pyramid.

The gallerys' lower walls will be higher than any man is tall.

Its upper walls will be seven courses of stone and three times higher.

The lowest of these seven courses will protrude beyond the lower wall by a palm width.

The next highest will protrude beyond the course below by the same measure.

This will continue through the seventh course.

The walls will form an arch, a corbeled arch.

They will not meet at the top of the gallery. The space between them will be roofed with limestone blocks.

Well below the south end of the gallery, a cavity will be cut completely through the stone of the ridge.

It will hold the middle chamber.

This chambers' ceiling blocks will shelter beneath the stone of the ridge.

As the structure rises, tier by tier, courses of this chambers' walls will enter the cavity, course by course.

Its angled ceiling stone will be levered into place, atop, peaked fillstone blocks. These blocks will lie atop the middle chambers' fillstone.

A horizontal corridor will run from the ascending corridor, along the concave wall of the ridge, then into the middle chamber.

On top of the ridge will be a leveled area.

It is here the high chamber will one day stand.

When the high chamber and the inclined gallery are both complete a short passageway, at the top of the great step, will run through the gallerys' south wall and into the high chamber.

Stonecutters will then remove the great step.

The entrance into the high chamber will then be closer to the gallerys', now lengthened, upslanted, floor.

Removal of the great step will be the removal of any hint of an underlying ridge.

The high chamber will be an eternal house fit for a pharaoh. It will be constructed entirely of granite.

It will be ten cubits wide, by twenty cubits long, by fifteen cubits high.[1]*

The height being from the top of its floorstone.

Above this well proportioned chamber, atop its smooth granite walls, will sit nine huge granite roofing beams.

Atop the roofing beams will sit four relieving chambers, also constructed of granite beams.

Unlike the smooth roofing beams these beams will be rough, except where they interface.

Relieving chambers are a part of the roof.

They will place enormous weight on the ends of the roofing beams. This will have the same effect as greatly increasing the length of each beam.

The rough granite beams will travel to their final resting place by other than ordinary means.

---

1 * One cubit equals twenty and two third inches.

The topmost relieving chamber will be roofed with great limestone blocks. These blocks will be emplaced east to west, along its length, and at an angle.

They will form a fifth chamber, with a peaked roof.

The blocks will be a part of the roof. They will separate the stone of the relieving chambers from the stone of the pyramid.

Now we move to events that will occur if it is the high chamber that will serve as pharaohs' eternal house.

A funeral procession will pass through the gallery and into the high chamber.

When all ceremony is ended, trusted stonemasons will open the high chambers' mystical passages.

They will seal and conceal the high chambers' entrance.

Only then will workmen be allowed to enter the gallery.

A false ceiling, made of wood, that the gallery wore for the funeral ceremony, will be removed. This will reveal a long line of plugstones, suspended by ropes, hanging along the gallerys' length.

The workmen will leave the gallery.

Stonecutters will enter the gallery, along with men of special crew.

The stonecutters will construct limestone benches.

The stone for the benches was stored in the descending corridor. It was stored beyond the entrance to the ascending corridor, out of the way of the funeral procession. Each irregular stone block will be large enough to construct two benches.

The men of special crew will line both sides of the gallery with benches.

The stonecutters will leave the gallery.

The men of special crew will place wooden planks atop the limestone benches.

Part of each bench, along its length, is not covered by the planks. It is the part closest to the gallerys' aisle. The empty space will be filled with basalt rampstones. Basalt is a harder stone than the granite of the plugstones.

The men of special crew will then leave the gallery.

Sailors of a now dead pharaoh will enter it, to render a final service.

They will lower the plugstones.

Wooden trigs will be inserted through holes in the wooden planks and into holes in the back of the limestone benches. They will protrude far enough into the aisle to hold each plugstone in place.

The upright section of each trig will be vertical to a bench, not to the ground.

The horizontal section will lie along a slot in a basalt rampstone.

When their work is done the sailors will leave the gallery.

Riggers will enter the gallery.

Wooden posts will be inserted through holes in the wooden planks and into holes in the back of the benches.

A post will be vertical to the ground, not to a bench.

Scaffolding will be erected.

The supports that upheld the plugstones will be disassembled and removed from the gallery.

The scaffolding will be disassembled and removed from the gallery.

The riggers will leave the gallery.

Two priests will enter the gallery.

Together they will raise each pair of trigs that hold back a plugstone. They will reinsert them in an out of the way position, in the same holes.

When the roar of descending plugstones has died away, the priests will move a limestone bench. Other priests, working from below, will cut through the gallerys' floor and enter the gallery.

A secret passageway, down to the gravel pit, will now be complete.

The unfinished low chamber was used to store the rubble from this passageway.

The secret passageway was not begun until late in the construction. It was not begun until it was known which direction it must take. Until it was known toward which tomb it must journey.

Everything but the limestone benches will be removed from the gallery.

The priests will leave the gallery.

After their return to the descending corridor a granite wall panel will be used to conceal the entrance into the steep passageway.

The next task will be to chisel away the limestone, at the mouth of the ascending corridor, that holds back the lead plugstone. This will be done by the trusted stonemasons.

When they are done the lead plugstone will rest on the floor of the descending corridor. It will be wedged between granite wall panels and the bedrock behind them. Other bedrock will be below it.

The great weight of the many plugstones above, some broken, and their uneven faces will make any upward quarrying, by tomb robbers, extremely hazardous.

Access to the lower section of the descending corridor, and into the heart of the pyramid, will now be barred; forever.

The one task remaining will be to seal and conceal the entrance into the pyramid.

My name is Dhan. I will be your guide.

You may wonder how it can be that a scribe pulled stone on a pyramid. That is the story I am about to tell.

My family was not poor. Nor were we rich, my father owned a farm large enough to support us.

However, I had a younger brother and if our father, upon his death, chose to divide the land between us, we would each raise a poor family.

While I was yet a boy some priests came to our village looking for youths with potential, for their temple.

A temple is really a town in itself. There are carpenters and metal workers, weavers and tanners, many skills and professions are needed and taught.

It happened not so swiftly, nor with such simplicity, but I was chosen for their temple. I had no love for farming, so both my younger brother and myself were overjoyed.

At the temple it was decided that I would be a scribe. I lived there and was schooled there for many happy years.

There I became a man.

Then came the day I was driven out, by a spider.

Which happened in the following way.

A demon, named Doubt, had taken possession of me. One by one my beliefs slipped away. My uncertainty grew. I began to lose faith. Incapable of answers I drifted, lost, on a sea of questions.

I carried my load of misery secretly. I did not seek the help of the priests. I did not wish to be stigmatized. I would wrestle the demon alone.

One day I was watching a spider build a web. There were young men at our temple studying to be engineers. This spider, I realized, had been born one.

This lowly spider had also been born with convictions, his instincts.

We both lived in the temple, he, in a heaven of certainty; myself, in a hell of doubt.

There is power in numbers. A similar power lies in letters, words. I was fearful it was through my love of words, of learning, that the demon had found a way inside me.

I decided to avoid mental work. I would live physical for a time.

Everyone knew of the construction that was underway at Giza. What better place to bury myself than in a necropolis?

That night I fled.

Now you and I must journey to a day when I had been at Giza for some time and no longer felt I was green.

The palms of my hands and the soles of my feet had toughened. Hard work had made me stronger.

I ate the same amount of onion and garlic that every worker enjoyed. I found they added zest to every kind of bean.

I was still a farmers' son. I had been accustomed to hard work from an early age.

While at the temple, no one there could wrestle me and win. But I had put the temple and my old life behind me and, I hoped, the demon.

I had a friend named Suri. He had been my mentor. Without his help I doubt I could have made the transition to life as a construction worker.

Suri, and others with whom I worked, must have realized that I had been schooled. They never remarked on it. They asked no questions.

I was surprised by their talent for minding their own business. I was also grateful.

Suri and I lived in Ragtown. We called the single mens' camp Ragtown because most of us lived beneath awnings of canvas.

Ragtown was clean. It was laid out as neatly as an army camp. Its rules were upheld.

Suri and I worked as laborers. Sometimes we toted water-bags or dry mortar. Other times we carried brick. Sometimes we carried sand, for the elevators.

Suri hated the work. He wanted to be a ropeman. Not a ropeman on the ground crew but a ropeman on the pyramid crew. If Suri wanted to become a ropeman I would also become a ropeman.

However, this was not something easily done. Ropemen earn more than laborers. We were paid in grain, which was easily bartered.

So Suri and I were not the only laborers seeking to become ropemen.

I now knew Suri long enough to guess there was another reason behind his ambition. Ropemen, especially pyramid ropemen, swagger. Suri, I felt, wanted to swagger.

Sometimes, when needed, laborers worked as ropemen. Suri and I had done so several times. Only a month ago we had worked as ropemen with the pyramid crew.

Today, as we sat and ate our midday meal, my attention was drawn to another group of workers, the riggers.

There was a ridge within the pyramids' perimeter. A side of this ridge had been worked on by stonecutters. Riggers were now removing the scaffolding that had supported them during this work. "Do riggers earn more than ropemen?" I asked Suri.

He nodded, while chewing.

"Yes", he finally answered, after swallowing. "It's not just knowing knots. It's working with the ropes and scaffolding." He shrugged. "You don't pick it all up in a week", he said.

"They're also part monkey", I said, gesturing toward the ridge with my leather water bottle.

They climbed ropes, walked a single plank over emptiness, busily removing and lowering scaffolding.

Suri nodded his agreement while looking toward the ridge.

"That's the work of thousands of years of wind and sand", I mused, out loud. I was considering the wearing of the ridge.

"And a few years of our stonecutters", Suri added, with a laugh.

The rams' horn began to blow. Our midday meal was ended.

Here I must tell of a farce that is enacted, every day, by pyramid workers.

All pretend ignorance of the fact that the structure will be used as a tomb. Internal structures, it is given out, have only religious significance.

An example of the extent to which this pretense is taken is the concourse.

A long concourse will connect the finished pyramid with a temple. It will be walled and roofed.

A funeral procession will pass within it. Few will know, other than those who are part of the ceremony.

When our workday is ended the pretense is dropped by almost all, except those who hope to become stonecutters. They will not take part in any forbidden conversation.

Nor will any stonecutter who yet harbors hope of becoming a stonemason.

It is the stonemasons, above all, who must remain tightlipped.

It is a handful of them who share in the deepest secrets of the structure, along with a handful of priests.

And along with a great many pyramid workers.

One result of all this pretense is that almost no ordinary citizen knows anything about a pyramid.

Another is that many know a good deal that is not true and, often, funny.

Sometimes, very funny, to the pyramid worker who overhears them relate it and, I imagine, to the pyramid worker who invented it.

Why do all go along with the farce?

Tomb robbers. It is understood that all this pretense frustrates tomb robbers.

It also frustrates those who believe turning a religious structure into a tomb is blasphemy.

Since I've come to Giza I now and then have this strange dream. As of yet I do not know its meaning.

In the dream, I am given a gift. Wondering and feeling anticipation I open the box that holds it. When the box is opened I find it is empty.

I always awake from this dream wondering if the demon yet lurks. Still, I do not call this strange dream a nightmare.

It was a rest day for Suri and myself.

In our land all work six days and then have two days of rest.

Here, at Giza, we do not all have the same rest days. We are divided into four groups. Three groups work while one group enjoys its rest days. Each group is off in turn. The work never stops.

A vinegar ration had been issued. I was soaking my feet in vinegar and water. One of our tent mates also happened to be off. He had joined Suri and myself.

By referring to our humble awnings as tents, I raise them up too high I fear.

"So" this tent mate was saying. "I hear you worked for Anan last month".

I nodded.

"For a week", I said. Suri let me tell the story.

It was the first time we worked for Anan. I was accustomed to foremen who expected we would do our work and left us to it.

Suri, myself and the others had not yet finished our first pull when Anan was among us. His face was red. The veins of his neck stood out thick as cords.

"You!" he jabbed a finger at one of our group, "and you!" he jabbed it at another. "You don't want to work! Take a walk!"

The two men stood, mouths agape. Anan gestured with his thumb, for emphasis.

"Take a walk!" he repeated. It was obvious he meant what he said. Both men took a walk.

This, I let our tent mate know, had been our introduction to Anan.

"That's him!", he said. "That's Anan! I worked for him when he was a foreman of laborers."

His look became thoughtful.

"I will tell both of you something", he continued; "as hot tempered as he is, if you work for him and you make an honest mistake, even a big one, he can just smile about it. He has no fear of what higher-ups may say".

"He is a strange one", he finished, shaking his head slowly.

I recalled those days we worked for him.

Anan never sitting, always standing. He never stood on one leg as some men do. Watching us as though weighing each of us.

I couldn't picture him smiling and said so.

14

"He smiles", our tent mate said. "he smiles the kind of smile a man usually hides behind his hand".

The man who spoke these words was an older worker, a bull of a man with graying hair. For the first time I noticed the light in his eyes. Some men have it. Usually, they can see more.

"Suppose" Suri joked, "it isn't an honest mistake one makes?"

"Then I would run", said our companion, seriously.

I laughed.

"Don't laugh", he admonished me, just as seriously.

I stopped laughing.

Later, when we alone, Suri said to me, "We never have any luck. If just once we could work for the right foreman...", his voice trailed off, he shook his head in frustration.

I had strained my vinegar and water through a cloth and into a jar, for future footbaths. I was on my way to rinse the cloth but waited.

I watched a boy going from dwelling to dwelling. He appeared to be one of those youngsters who earned their living running errands.

He left a nearby dwelling and approached our own.

"Does Suri and Dhan live here?" he asked, upon arrival.

"Who is it that wants to know?" Suri asked.

The boy thought for a bit.

"I have a message for them", he said.

"And what is that message?" Suri asked.

"You can speak your message", I told the boy. "He is Suri and I am Dhan".

"You are both to move your belongings to that part of camp where the ropemen live", he said. "I am to show you where you will stay."

I watched Suri pale. I realized why. This seemed just the sort of prank that workers played on one another.

"If this is a prank", Suri slowly said, "you will be running your next errand on one leg."

The boy strained to make himself taller.

"This is no prank", he said, indignant. 'Anan has sent me!"
So that is how Suri and I became ropemen; ropemen on the pyramid crew.

The Second Scroll

The east road runs from the quarry to the east ramps. All of the stone used in the construction of a pyramid, and more, will climb the east ramps.

The east road also leads to a stoneyard.

It is here that stonemasons shape many of the limestone pieces needed for the special construction. The base of each piece will be its skid.

The great river wends its way north, past the construction, then makes a bend, east and west, before continuing north once again.

The docks lie north and northeast of the construction.

Here the white limestone blocks, for the pyramids' outer coat, will arrive.

This stone will be delivered to the stoneyard by way of the north road. The base of each casing stone will be its skid.

One day each white stone will be toppled. Another day each former base will be cut away, at an angle.

A line of this white stone will lay along the pyramid the same way it once laid in the ground. In this way each white stone will fit, side to side, with little effort.

Here the high chambers' granite floorstone and wall-stone, and the granite plugstones, will arrive. All this stone will be delivered to the stoneyard by way of the north road.

Here the many rough granite beams of the relieving chambers will be off loaded.

Here the nine great roofing beams, for the high chamber, will be landed.

All of these beams will be delivered directly to the structure by way of the wide concourse road.

They will climb the north ramps.

Now we are ready for a trip back to that time when the greatest of pyramids was being constructed.

Now we stand on the very edge of an upper tier of that structure.

Ah! The magic in papyrus.

You're surprised. You expected a different pyramid.

You must be patient. That will take time.

What you see is what the early magicians created, a step pyramid. The pyramid you thought to see is inside it, like a caterpillar in its cocoon.

Once, long ago, only square or oblong tombs were built. They were made of brick.

Why such tombs?

A creation story is told. In one version a legendary bird alights on an island in the waters of chaos. With a single cry it ushers in coherence and all its underlying causes.

Prior to that cry there was no such thing as coherence, or cause.

These tombs represented that island, the island of creation.

The nomads who roam our land refer to these tombs, in their argot, as mastabas. Some say it is their word for a bench, blocks of stone placed outdoors for use as resting places.

When stone began to be used, to make such structures, men thought to stack them, to create more impressive structures. They built step pyramids.

We then adopted the nomads word, mastaba, to describe a single step on such a structure.

To the outside of each mastaba are its working lanes.

Alongside these are the ramps, ending in level sections, called roadways, that run alongside the working lanes.

The width of a ramp or its roadway is equal to the width of a working lane.

The bottommost ramps of the great step pyramid fill one half of the cleared area around the great pyramid.

Emptyness fills the outermost half.

During the first phase of construction a great deal of time will be saved by using corestone in the inner pyramid.

Large islands, composed of corestone, will be created; islands as high as a given number of tiers.

Much of the corestone will have cuts so that it can be fractured into tierstone, using wooden wedges.

Other times whole blocks of corestone will be toppled, to create tierstone.

Tierstone connects the innermost construction to the ramps. It is needed around internal construction, to facilitate that construction.

When tierstone reaches the height of the corestone new islands of corestone will be created.

Rubble will be taken down the west ramp.

Now you must learn some basics before coping with the pyramids' deeper mysteries.

I do not mean the arithmetic involved in building a pyramid. I know little of such things as the ratio of bases to heights.

What I know about building a pyramid I learned while building one.

I know about being awakened by leg cramps those first weeks, until ones' legs became accustomed to the work.

I know other things as well.

Work is passed along, from gang to gang.

Five teams pull stone along four sections of road. When the sixth stone is being delivered to them, the first team will already be there to receive it.

In this way all of the stone is kept in motion.

At times gangs will be joined together, into larger ones.

At times gangs will be broken up, into smaller ones.

The men of chalkstone and slatestone, the foremen and the general foremen, will meet to plan such moves.

How is work put onto a ramp?

Is it to be pulled on, by brute force, by a host of ropemen; with the front end of the work grinding its way up the ramp, until enough daylight shows beneath it that it breaks in two?

No. To put work onto a ramp an incline is used.

An incline is the continuation of a ramp into the ground. It extends to the side of a ramp.

At its deepest an incline is one cubit deep.

When a stone is pushed down an incline it will be on the same angle as the ramp. It can then be levered along the incline and onto the ramp.

All ramps are six to one. This means that for every six measures of length there will be one measure of rise.

The workers deem this so. Those above us have their duties and their rights, we too have ours.

Building stone travels from the low ground of the quarry to the structure, standing on end.

It is levered into the bottommost east incline.

It is moved along the incline to the base of the ramp, then up the ramp.

The east incline will not be as long as the cleared area is wide. This way stone can be moved to the inclines far side.

At the top of a ramp a stone falls onto planks that protrude from grooves in the roadway. The rock beneath these planks is greased with tallow.

Stone and underlying planks will be levered into the working lane.

Men, using pushsleds, push the stone along the working lane and down the incline at its end.

Now let me take you to see the crazies that put all of this stone into place.

They eat a lot of garlic. It gives their sweat a strong odor. I hope it will not offend.

Notice the planks. Carpenters fasten them down to the stone of the tier. They're greased, using sticks of tallow. On the working tier such planks are a big help in moving stone.

While we walk I will talk.

There is a surveyor, named Aneferu, who is fond of pointing to some huge block of stone and asking some young worker if he knows how it will be put into place.

When that worker replies that he doesn't know, it makes Aneferu very happy.

Now he can say "Slowly, very slowly", and walk away laughing at this ancient joke.

Often, others nearby cannot help laughing too. Not at the ancient joke but at Aneferu's continual amusement over it.

For the internal construction we use limestone skids.

Most often a skid is one with its load.

The men of special crew do all the internal construction. They do it one course at a time. All is planned in advance. There is no need to improvise. No one likes surprises.

Here we are, the crazy house!

Two men swing a short beam affixed to ropes. They're pounding a stone into place.

Each time the beam hits the stone one man grunts the same obscenity.

His partner no longer hears him.

He hears only the sound of wood on stone.

Somewhere a voice yells "Engineer! Get your fat ass out of the way!"

The eternal cry of kitchen workers "Hot stuff! Hot stuff!" is borrowed by a worker guiding a fast moving stone into the area.

Ignoring all of this, carpenters reposition greased planks.

A grinning worker passes close by and you smell the beer on his breath.

In the midst of all this, contented, stands their foreman. He wears a paternal smile on his face. "These are my lunatics" it seems to say.

21

He has good reason to be contented. For every evening the fat-assed engineer must report to his superiors that the work goes well.

He wonders how this can be? He feels this is not the way a pyramid should be built.

He is wrong. Pyramids have always been built this way.

Tonight I walk out to the structure. I often do. I often spend part of my rest days there.

I will put a blanket over my shoulders. You, I know, will not feel the deserts' cold.

Tonight I visit the stoneyard.

While we walk I will tell you a story.

When I was a young boy a juggler entertained some people in a house in our village. A juggler is someone who pretends to work feats of magic. Someone who seems able to pull objects out of the empty air.

In the darkness I stole up to a window of that house, to watch the show.

The juggler began by showing his audience a wooden ball. The audience watched as he put the ball into his mouth. The ball swelled out one of his cheeks. He shifted it to the other cheek, which swelled out in its turn, then back again.

Finally, with convincing facial contortions and to sounds of astonishment from his audience, he swallowed the ball.

All was well however. In another moment he approached a member of the audience and retrieved the same ball from behind his ear.

I can vouch for it being the same ball. My window onto the show was behind the juggler. From that angle the secret of the whole illusion had been revealed to me.

It may have been that very night that I lost my faith in magic.

It was a long time before I realized I had seen real magic at work that night.

The magic was not in the jugglers' sleight of hand. Nor was there any magic in his acting ability. The real magic, in all probability, did not lie with him at all.

The real magic lay in the creation of the whole illusion. Even in the creation of a humble illusion there is the real magic of creation. An ability that is only ours', and heavens'.

Here is what I wished you to see.

This multitude of stone is more impressive in the dimness of starlight. These are stones for the inclined gallery.

All of this stone is gathered here because, once, one man stood where we now stand and heard music.

He heard the music of work songs that men sing as they lever heavy loads. There is no denying the magic in music.

He also heard the music of numbers.

These stones will advance at different times. They will rise to a variety of levels. They will come together at various times and in various combinations.

These stones dance to a music, the music of numbers.

Though unheard, that music is here. Everything dances to it. This whole structure is becoming permeated with the magic in that music.

Now, through another magic, the magic in papyrus, we move to another day. A day when Suri and I were no longer laborers.

We began work as ropemen much further down the tiers. Those first stones we pulled had been covered over by many others.

This day our team worked a top pull on one of the ramps.

It was late in the day. It was very hot. It was one of those days when the air felt almost liquid.

"What do the quarrymen have against us?" Suri asked me.

He expected no answer. I was too tired to respond. We were all tired. He was referring to our next stone. Our next pull, in ropemens' language.

It was a monster stone. The stones had been a uniform size all day. Why this outsized stone I wondered?

We unloosened our towrope, knowing the monster stone would be next. We could see it moving toward us.

"Do you think it's hot up here because we're so much closer to the sun?" Abi jokingly asked his cousin Amus, as he wiped at his sweat.

My own mouth was dry and I had finished my water.

Amus didn't show any sign he had heard. He was gazing down at the ground crew "rushing rock", as they called their work.

"I couldn't do their job", Amus said. He used his chin to indicate the entire ground crew.

"Pulling work on that greasy soapstone! That job's a slice of watermelon", exclaimed Abi.

I looked down at the ground crew hurrying with stone. Then I looked at the monster stone. Right then their job didn't seem so bad at all.

"But it's work stopping those stones", Amus said defensively, as we neared the monster stone.

I saw Seti looking at Amus with one eye closed. Next to Nin, our team leader and an ex-soldier, Seti was the eldest of our team.

By now we had reached the monster stone. Nin and Suri were attaching our towrope.

"The way they rush back for another stone", Amus continued, "we can take our time."

Abi just shook his head.

"Amus?" asked Seti.

"Yes!", Amus answered.

"Shut up", said Seti.

Amus let out a burst of laughter. It was his "you walked right into that one" laugh.

Seti looked at me a bit sheepishly as we both laid hold of the rope.

"He fooled me too", I told him.

"You cucumber!" Abi said to a grinning Amus, as they both laid hold of the rope.

"Lay hold!" Nin bellowed.

"Heave!" we all responded, and did.

When next my right foot dug in Seti yelled "Heave!' then Suri in his turn "Heave! Then myself "Heave!", followed by Abi, then Amus; even the twins, Tu and Ty, working the levers, sang out.

It was a thing we sometimes did, without planning, the way a team sometimes does a thing. We sang out in order of age and, somewhat, in order of experience.

As we drew near the top of the ramp I wondered, would we get this monster up and over the edge?

I didn't think so. I felt something akin to panic.

"Pull!" I yelled.

The monster was starting its climb above the edge.

"Pull! Pull!" I yelled louder.

The stone climbed higher; a little higher. I continued my yelling.

"Pull! Pull!" louder and louder.

The monster stone continued its slow rise above the edge, hesitated, then fell slowly; banging down onto the planks embedded in the roadway.

I had never stopped my yelling.

Amus was having a fit of laughter. Abi joined it. Even my friend Suri let out a laugh.

I felt embarrassed.

"It all helps", Seti said to me, "don't let their laughter bother you."

The rams' horn began to blow, signaling the end of the workday.

"Let's go eat", Nin said.

"And get water", Tu said.

"Lots of water", said Ty.

As we walked back to camp I recalled I once thought that ropemen swaggered.

Ropemen don't swagger. They walk the way they do because they know how a pyramid is built. A pyramid is built pull by pull.

Three pyramids were planned.

Only one will be completed.

No one knows which one.

Four roads will connect all three structures.

They will connect them at their corners.

An earlier structure will be enveloped by a later one, just as the ridge, and the mount it was on, would be enveloped by the low pyramid.

Atop the second mastaba of the low step pyramid the middle chamber will be roofed.

Atop the second mastaba of the middle step pyramid elevator compartments will be constructed.

There is good reason for elevators.

Construction of the relieving chambers will occur course by course, as the pyramid is rising all around them, tier by tier.

It was planned to roof the high chamber atop the second mastaba of the high step pyramid.

Each thing was planned to occur at one third the height of a structure; an outer structure. A step pyramid that will never be completed.

To construct a low pyramid a low step pyramid must also be built.

Were the low pyramid completed few would know the ascending corridor existed.

The upper section of the ascending corridor could not have been connected to the lower section.

There could be no lower section. The low pyramid would not be wide enough to contain it.

Plugstones would seal the low pyramid from ground level, by blocking the descending corridor.

An entrance into the middle chambers' corridor lies along the ascending one.

A day will come when that middle chamber corridor leads to a roofless middle chamber.

The tops of its north and south walls will be level with the tops of its east and west walls.

The tops of all four walls will be level with the top of a mastaba.

The middle chamber will be filled with limestone blocks.

The chamber will also hold a large statue, safely in a niche, and a sarcophagus.

The middle chambers' ceiling stone will climb the east ramps.

Imagine a large rectangular stone.

Imagine one corner, a lower corner of that stone, cut in the shape of an angled ceiling block.

Two lines of such stones, emplaced over the middle chamber and arched toward each other, will form the chambers' ceiling.

From below, the ceiling blocks will appear cantilevered.

There is power in shapes and forms.

If one probed behind the wall, with a thin strip of material, they would be confirmed in that belief because of how the inner wallstone is shaped.

One day, before a funeral, a visiting dignitary from a puppet state may look on the middle chambers' wallstone and smile. He may even smirk.

The courses appear uneven.

By such devices will an onlookers eye be led astray. He will not see beyond appearances. He will not perceive the true shape of the ceiling blocks.

When the last ceiling blocks are emplaced fillstone will be removed from the middle chamber. Circumstances will not allow for all that fillstones removal. When the decision is made to construct the high pyramid, removal of the middle chambers' fillstone will be curtailed.

Were the middle chamber to be used as an eternal house entry must be made into that chambers' corridor.

After the funeral, men would take stone from the south end of that corridor and use it to plug the entrance, at the north end.

This would complete a roadway for descending plugstones.

One day a milestone will be reached in the construction.

On that day men will enter a tunnel. It leads toward the middle chamber. They will cut through its east wall and enter that chamber.

Some will retrieve the statue but not the sarcophagus.

The sarcophagus will lay beyond their reach and beneath fillstone.

Others will take stone from the floor, at the south end of the middle chamber corridor, and use it to plug the entrance at the north end.

The middle chamber will be abandoned.

Its mystical passages will remain closed.

The tunnel will be filled. Its outer entrance will be covered over by the stone of the high pyramid.

Now there will be no more uncertainty.

Now there will only be the high pyramid, the greatest of pyramids, the great pyramid.

Being a scribe means more than having the ability to write. Scribes are educated in many areas.

At the temple it was decided I would study foreign lands. I would learn the strange beliefs and customs of other people.

Which led to my making a discovery.

In these diverse lands are many who share a common belief.

People who live great distances apart, who speak different languages, share this same belief.

They believe that, once, men dwelt together in one perfect place. This is written in different holy books.

I also came to accept it as true.

Why?

I cannot say. It was not a belief I came to through reason. So there are no reasons to give.

I felt it was true.

This long ago place must have been one of perfect harmony. Such harmony could come from all sharing common beliefs. You might say from all having a common ka, or nature.

This is not the ba, or soul.
The ba is a piece of infinity, a spark of divinity.
That spark is dim in some, burns bright in others.
If you look into the eyes of one hawk you see the ka.
You see the nature common to all hawks.
The same is true if you look into the eyes of a cow.
Men no longer have a common nature. Men are different, one from another.
However, I now accepted that, once, men must have had a common nature.
Like other creatures we would have been born with common beliefs, with common answers, with our own instincts.
Next I must tell you about the dragon. A monster you no doubt heard of, a beast you may have seen depicted.
In our land, in the house of truth, are the ancient bones and blackened teeth of one of these beasts. Teeth blackened by its own fiery breath some say.
I have handled these bones and teeth. They are no jugglers' trick. Some of the fangs are much larger than a mans' forearm. You can imagine the size of the bones.
What change, what misfortune or series of misfortunes, occurred to bring about this monsters' passing?
These same changes may have brought an end to our paradise on earth.
Such changes would bring uncertainty and doubt.
We learned to doubt.
The demon employed it zealously.
It became a new instinct.
So it was not just external changes that destroyed our ancient world but internal ones as well.
We, who had once been born with all the answers, with instincts, were now born with only questions.
Our instinct to create any specific thing would also be undermined.
We were left with a general urge, to create.
We were left with an innate need, to create.
We were left with a new instinct, to create.

This is how heaven gave us imagination.

Instincts are heaven sent, beliefs are manmade.

Now I understand why men search after knowledge. I had been one of the searchers.

It is to fill the emptiness left by the loss of our ancient instincts.

Yet I also know that what one man proves, another man will later disprove.

What men of one age hold to be true, will amuse men of a later age.

The box of knowledge can never be filled because it is always being emptied. I know this because the demon himself has told me.

I also know it is true.

Since I've come to Giza I've come to realize that doubt is but a new instinct, sent by heaven to destroy our more ancient ones.

But that demon will no more allow us new beliefs than heaven allowed us our ancient ones.

No demon had taken possession of me. It was born in me. This is true of all men.

As I learned, so did he.

He grew shrewder.

His questions became slyer.

I too am growing shrewder.

I have mentioned my ability as a wrestler.

Besides strength a good wrestler needs determination.

I said I would wrestle the demon alone.

It will be here, at Giza.

This is the chosen ground.

The Third Scroll

Now let us move on to the magic in pictures.

In our land we often use pictures to tell a story.

Changing pictures, shown upon a wall. Sometimes the easiest way to see a thing is with your eyes.

I will try to use words to create a series of such pictures, to help me tell a number of stories. Taken together they will be but a single story, the story of our team.

Years ago, when work first began on the pyramid, the workers' flour was contaminated with sand. Unscrupulous contractors often did such things to increase their profits.

There had been no response to their complaints so the men took matters into their own hands.

The result was a strike.

Seti had been one of the organizers of that strike. The man he had supported, as its leader, proved more than capable.

The men held out for three days. They were ready to hold out longer but it wasn't necessary. Their one demand was met. The contractor was replaced.

On the right side of the picture stands a large group of men. They each have one hand raised, with the palm forward, toward two men on the left.

These two men are drawn larger, to show their authority. One has his arms extended, toward the group on the right. The palms are up, in entreaty.

The other has a stick raised over his head

Among the men on the right is one man also drawn larger. He looks at the two men on the left without expression. His arms are crossed

Only weeks after the strike ended this man was made foreman. Foreman among those who serve at the desire of the royal house. A rare honor for one without influential friends, one such as Anan.

Seti remained who he was, a man sensitive to any abuse or injustice. He also remained what he was, a ropeman, much to his fathers' disgust.

Setis' father also worked on the structure. He was a stonecutter. He had made many friends during his working years.

He knew some of these friends could help his son. He wanted Seti to be more than a ropeman.

Seti was a widower. His wife and daughter had died of the same illness that he had survived. So there were no loved ones whose lot he might improve.

He had no interest in his fathers' plans. He would remain where he felt he belonged.

Next I will tell you the story of the twins, Tu and Ty. I will not need to be asked questions, as they were, to help me relate it. Nor shall I tell it in bits and pieces, as they did.

I shall relate their story with greater ease than they knew, living it.

The twins had only their father. Their mother had died giving them life. He was part owner of a ship that plied our great river. He was also a gambler.

He and the twins lived and worked aboard the ship. When he died, quite unexpectedly, his debts about equaled in value his share in the ship. All that remained for the twins as their estate, so to speak, was passage to Giza.

Here the twins spent many of their rest days along the banks of the river they loved. They sought work aboard any ship. They spoke to captains who had known their father and to captains who had not; without success.

Each of the next pictures show the twins standing before a sailor in the garb of a captain. In each, the captains' arms are extended out to the sides, the palms are up, the shoulders raised. He looks past the twins.

There are many such pictures.

The river carried shiploads of grain, bound for foreign ports, north. It bore loads of cedar and other foreign wood, south.

Cargos of papyrus and linen flowed north, with the current. Ivory and olive oil sailed south, against it.

It hauled the stone of our quarries, granite and limestone, alabaster and turquoise.

Copper and gold, from the mines of the eastern desert, were part of the flow.

The river trafficked in wealth.

Many men, too many, sought a place on the great river. Almost always, when such a place became available, a relative waited to fill it.

There will be more, concerning the twins, but it must wait.

Before I begin the next story I must first tell you that commerce in our land is made easy by brokers.

If a man wishes to sell a house he goes to a broker. What goods the house is bartered for matters not. From the broker the seller receives payment in whatever he wishes, copper or gold, perhaps grain.

The grain in our granaries is traded many times before ever leaving them.

Abis' father was a broker but Abi had no love for trade, Abi loved the soil.

Amus did have a love for trade. He, therefore, became the one to assist his uncle in the family business.

Which led to Amus and Abi conducting a small business of their own, on the side. They did trades that Abis' father deemed not worthy of his time.

Such a situation could not remain secret for long, nor did it. With discovery came loud recriminations, from both fathers.

In Abis' case these went on for too long.

Now you see a series of three pictures.

In the first, two older men berate two younger ones.

In the second, only one of the older men berates one of the younger ones.

The third picture seems to be the same as the second but if you look out the window, in the picture, you see that the tree in the garden now bears fruit.

Abi decided to leave home.

Ready for further adventure Amus decided to join his cousin.

A fourth picture shows two young men on a road. They carry bags, like those used by travelers, on their shoulders.

In the distance is a partly constructed pyramid.

Abi was fascinated by the earths' power as a womb. At Giza he spent many a rest day helping farmers, he had come to know, work their land.

Enjoying a home cooked meal, in the company of those who enjoyed his company, gave pleasure to Abis' life at Giza.

Amus spent many a rest day in the shop of an elderly broker named Pooth. Amus listened and learned. He learned well.

With Pooths' approval he joins the twins along the river. He is also looking for a captain. One who has the space to bring deck cargo to town.

In the picture you can see Amus has found his man. You can tell by the way he and the captain shake hands.

You can also tell by the identical grins on the faces of Tu and Ty.

Pooth will pay over a sum of gold to the captain. On his return to Giza the captain will bring a deck cargo of wine. A commodity Pooth feels can be easily and profitably exchanged.

Nin, whose story must wait, and Seti were both known and respected. Nin by other team leaders, with whom he spent a good deal of time, and Seti by almost all workers.

They would arrange for the honest labor of unloading and warehousing the cargo. The honest labor of other

teams. Amus was insistent on this. Our team would only provide the rest day labor of others.

Abi would arrange for the oxcarts and drivers.

I would take care of all documents.

Would Amus make us rich?

Yes, among pyramid workers. I doubt even the farmers whose oxcarts we rented would envy our share in each
venture.

However, there were other riches.

The twins grinned more often. They knew they had
played a role in launching the venture.

Today was a rest day. My destination was the inclined
gallery being constructed by the men of special crew.

I had filled a water bag with cold water. Water is always
welcome to those who toil in the desert sun. It would make
me more so.

It didn't take long to arrive amid the shouts and the song.
As I approached the group I had chosen to visit there came
a wave of the hand from the group leader. Others called
out greetings and I received a nod of forbearance from
their foreman.

I handed the group leader the water bag and found
a seat on a stone, far enough from the work to satisfy the
foreman.

It was really a seat on a folded loincloth, today I wore a
tunic.

The walk with the water bag was a cheap price to pay
for my seat. The beauty of the work songs sung by special
crew equaled that of any chanties sung by our sailors.

These men do all the internal construction. All are older hands. Men who have worked on construction up and
down the river.

This is also a favorite expression of theirs. If one of them is
told a tall tale he's "been up and down the river too many
times to believe that story."

Before the first stone of the low pyramid had been laid
stonemasons were fashioning the stones of the gallerys'

long east and west walls, like much of the construction these stones were shaped to fit together.

All of the gallerys' building blocks are stones that are stacked vertically. This is the only way the gallery will stand, unchanged, forever.

When the gallery is finished and someone stands within, they will believe themselves to be standing between sloped walls.

Were its wallstone truly laid on an angle the gallery would slide into oblivion. Compression would destroy it.

Were it another time I would take you out onto the stepped ridge and show you a wallstone sitting, angled, along the slope that is to be the gallerys' aisle.

Immediately behind it, sitting on a level step, is a level block that is partly obscured by the one on the slope.

We would then walk toward the opposite side, in order to see how much of the block on the slope is obscured by the one on the step.

Only then would I tell you they are one and the same block.

There is power in shapes and forms.

Limestone block by limestone block a stone scaffold will be raised down the center of the gallery.

Stone, arched over the lower entry into the gallery, will be a part of the stone scaffold.

Space will be left to either side of the stone scaffold so that men can enter and exit the gallery.

As the gallerys' north wall rises the stone scaffold will rise.

After the high chambers' floorstone is emplaced the plugstones will be emplaced atop the stone scaffold.

One day stonecutters will reduce the stone scaffold to rubble.

The rubble will be removed from the gallery by way of the descending corridor.

The double benchstones that line the lower descending corridor will be created from the stone of the inclined gallerys' stone scaffold.

As that stone scaffold is reduced each, upright, double benchstone will be laid down, then sent down the ascending corridor to the descending one, finally down the descending corridor to its assigned waiting place.

This will be done after the secret passageway is nearly completed.

After the funeral, when the benchstones are returned to the gallery, the secret passageway will be completed.

The entrance to the descending corridor is on the north face.

In the low pyramid that entrance is at ground level.

In following pyramids it rises above ground level.

In the high pyramid that entrance will be thirty-two cubits above ground level.

In the high pyramid it is offset, to the east of center, by fourteen cubits.

The first plugstones will enter the gallery by way of the great step.

Each stone will be individually roped, at its thick upper end.

Each will be levered down the stone scaffold.

Sailors of the pharaoh will await each stone.

Above the sailors heads is a round beam.

Each beam is horizontal, parallel to the ground not to the gallerys' floor.

Each beam lies atop two wedges, one at either end.

Each wedge rests between a pair of upright wall beams.

The back of each wall beam conforms to the shape of the wall it presses against.

Their faces conform to the shape of the wedge.

It is the weight of the circular beams and the wedges that keep the wall beams pressed into the walls.

The sailors plan to add the weight of a plugstone to each round beam along the gallery.

They wrap turns around a circular beam with the free ends of ropes attached to each plugstone.

In time plugstones will hang, suspended, along the gallerys' length. They will hang above a false ceiling, waiting, until after the funeral ceremony.

Below the false ceiling will run the funeral corridor, four cubits high, by four cubits wide.

I fell into a reverie.

Sound, already muted in the heavy air, seemed more so. I half-dreamed of new made man, without his old instincts. Having to learn how to think, or having to learn how to avoid it.

One such man arises each morning in the ashes of his old paradise. He is a gatherer. He gathers shells and colored stones, then he sorts them. He fashions them into ornaments he calls knowledge.

He believes that one day, when he possesses enough of these ornaments, he will attain understanding. Then he will live in a new paradise, better than the one he lost.

Another new man also carries the new burden, doubt, in his breast. There also beats his heart. This is the man the demon fears. This is the man doubt has set free.

This man uses the rubble of his old prison to build a workshop. Here he keeps doubt chained. Nor can doubt escape, anymore then men can escape doubt.

This man encourages doubts' questions. He seeks answers. He makes no ornaments. He creates. This man is the magician.

I come fully awake. Nearly falling off of my stone. I rub my face.

Over time, as I watched the construction unfold, I came to realize just how a pyramid is created. I came to realize it is created by magic. The same way a poem is created. It springs, magically, from the imagination.

It took lifetimes to create this pyramid.

First men created lesser structures. To create even the least of these, solutions had to be found to problems. Each solution a miracle of invention by a man of imagination, a magician.

Pyramids are not raised up by men with a knowledge of numbers but by men with a gift for creation. Their miracles become the shells and colored stones other men gather.

It is the makers of miracles, the magicians, who made this pyramid possible. It is the gatherers of shells and colored stones who are in charge of the construction.

Tonight we will walk out to the unfinished pyramid.

You will see a finished one, complete and wearing a coat of white. I speak of the capstone.

Many regard the capstone as the pyramid. All else they view as pedestal.

Where I am going is close by the sacred stone. You will have a good look at it. It must be the first stone onto the site.

It must be the first stone onto a new tier.

No other stone may ever rise above it.

The stones awaiting its arrival, that first day, are lower and are not within the perimeter.

The sacred stone, riding atop a painted and gilded sled pulled by priests, makes its way through a joyous throng.

There are many happy women in this crowd. Women who know their husbands will have work and their children will be fed for years to come

The sacred stone is levered off of the sled.

It is levered onto a rampstone, to the sound of cheering.

The sled is removed.

Another stone is pulled into the area. Its top has also been cut into a ramp.

The low end of this rampstone is maneuvered under the angled capstone.

Other such stones follow, each increasingly higher.

Before long the capstone will sit, over one tier high. It must await the laying down of the first tier of the low step pyramid.

It will climb through a mastaba.

It will be levered southward,up new rampstones, then backward with its rampstones, while a mastaba rises all around it.

The upper part of the capstone is a pyramid, symbol of the sun cult.

The lower part is a square stone slab, the island of creation.

The pyramid sits well within the slab.

On another day, at a much greater height, most of a new tier will be laid in front of the capstone and its rampstones.

Men, atop the island of creation, will use their own weight to lower the capstone onto this incomplete tier of stone.

Riggers have wrapped the island of creation in rope. Men, using these ropes, will pull the capstone onto the final tier.

Men now complete the laying down of that final tier of stone.

After the tier is completed the capstone will be centered.

After the capstone is centered, stonemasons will give it a final shaping.

This is as far as I go. I do not care to get too close. Many young priests take the job of guarding the sacred stone very seriously.

You can see it gleaming white in the moonlight. It stands taller than most men.

Here, gleaming black in that same moonlight, is what I have come to see.

These are the rough granite beams. They lie atop sand, mostly above their brick lined pits.

A granite beam is rocked from either side, on its rounded bottom, by men using levers. Other men pour sand. The continual rocking compacts the sand beneath the beam and the beam rises.

To the south of the ridge is a row of elevator compartments.

Each of these compartments contain two of the thinner rough beams, lying lengthwise, east to west.

There are eight such compartments.

At the proper height these beams will mount the high chambers' roofing beams, or a relieving chamber, in pairs.

They will lie east to west, lengthwise, four in each chamber, two to either side.

Remaining beams will be moved forward. Rearmost elevator compartments will be closed first.

Arranged to east and west of where relieving chambers will one day stand are the elevator compartments holding the thicker, rough granite beams.

A row of thicker beams will lie atop the thinner ones, lengthwise, north to south.

This arrangement of elevator compartments is possible because there was enough room atop the structure at the height of the second mastaba of the middle step pyramid.

Look closely. Notice how smooth the rounded bottom of this rough granite beam is becoming.

They're all like this.

It's from the constant rocking on top of the sand.

If I asked how such a beam could be lifted to a great height, could you have found an answer?

Would you have realized an answer lay all around you?

Would you have sifted sand through your fingers as you pondered?

Some time ago I realized I was possessed by a demon.

Then I realized he had come to torment all men, to acquaint us with endless doubt. Heaven, however, had given us a gift, the gift of imagination.

It provides unending answers, even to the now endless questions.

Now I must return to Ragtown.

Amus and 1 have an appointment tomorrow, a long ago tomorrow.

"Here he is!" yelled Pooth at Amus, as we entered his shop the next day.

"The dealer in words!" cackling.

"My honorary nephew!" more cackling.

"And he has brought the ravisher of women in the house of truth!" louder cackling.

The last was directed at me. Pooth moved my schooling, and any depradations he concocted, to the more exalted house of truth. It increased his amusement.

"I ravished only those whose protests 1 thought false", 1 said to Pooth.

"That would keep you busy! There'd be little time for lessons!" Pooth replied.

"Greetings dear uncle", said Amus, ensnaring Pooth in his own jest.

Old Pooth shrieked in glee.

"Has the world ever seen such gall? He robbed his own uncle of all he owned! Left him poor in his old age! Now he will adopt me!"

Through all of this Pooth occasionally had to dab at tears of humor with an expensive sleeve.

After all was signed 1 left Amus to listen to stories of long ago commercial machinations. I removed myself to a small park by the river. Which is how I came to meet Nebri.

"Do you mind if I sit here?" he asked me.

A glance around showed that other stones, set under the trees by the river, were filled with people.

"If you don't mind sitting next to someone possessed by a demon", I said to the stout man in the costly tunic, who stood by my bench. He hesitated but sat.

"Are you truly possessed?" he asked. He looked at me, then looked quickly away.

"We all are", 1 informed him.

"Including myself?" he asked the river.

"Especially you", I told his expensive tunic.

"How could such a thing have happened without my knowing?" he asked wryly.

"It happened long ago", I informed him; "before gold became valuable there was a golden age. We ate only manna. We were very happy. We even knew right from wrong," I said. Like a spider I once knew, I thought.

"What changed all this?" he asked. He chanced a smile, a small one.

Change, I told him, was the first law of nature. God, I told him, does not just create life but goes on, reinventing it.

He knew of the dragon. He had been to the house of truth and seen its remains.

I spoke of heaven sending uncertainty, or events that caused uncertainty, into the world. I explained how we learned to doubt.

I told him how our desire to create the old things became abridged, became simply a desire to create. How that became another new instinct, imagination.

I explained that these new instincts, doubt and imagination, led us to freedom, to a soul, and to the burden of immortality.

Before Nebri left, he told me his name was Nebri, he invited me to his home. He described his house and told me how to get there. He told me I could bring a friend.

He knew I worked on the pyramid. I had told him I was a ropeman.

In our conversation I had also admitted to some schooling.

I sat and watched the river.

Doubt, I knew, brought only questions.

With them, he could cause despair. Such hopelessness that a mans' own hand often became the weapon raised against him.

Doubt is laid to rest by finding answers, by imagination.

Soon, the answers are under fresh attack by doubt.

Doubt without imagination is unimaginable. They are like a left and right sandal.

Is man an instinct in god?

It seems inevitable that he would raise a creature above other creatures. A creature who could create. A creature who would create him, knowing his existence answers our larger questions, our larger doubts.

He created the spider to create a web.

He created us to create him.

Our creation is as necessary to our own existence as the spiders' is to his, and is just as real.

Why must he prove his own existence?

He created us to do this.

He gave us the tools, imagination and doubt.

Who has not heard of the first cause uncaused, the cause of all causes?

God and we create one another, as intended.

I look upon this as I look upon the flow of this great river.

That is why I ask, is man an instinct in god?

The Fourth Scroll

The next piece of construction to be considered will be the construction of the high chamber.

Once the ridge and the great step are reduced to the proper height the floorstone of the high chamber will be laid.

The rough beams will already have begun their rise. They will be arrayed around the floorstone, proving the worth of elevators.

As courses of the high chambers' walls rise they will be held firmly in place by blocks of limestone fill. Such blocks will fill the high chamber.

Some later plugstones will reach their place in the gallery by traveling over the high chambers' fill, then down the gallerys', ever rising, stone scaffold.

This fill would also provide support for the high chambers' roofing beams as they are levered into place and until relieving chambers are constructed.

It was planned that when the tops of the high chambers' walls were level with the top of the high step pyramids' second mastaba, the great roofing beams would be emplaced.

There would be enough room, between the high chambers' west wall and the west elevators, to allow for the emplacement of the final roofing beams.

The rough beams of the relieving chambers would continue their rise, along with the next mastaba.

Once emplaced, a rough granite beam that needed to be upheld would be upheld by a limestone block beneath it, until it was upheld by other means.

One day these temporary supporting blocks will be reduced to rubble.

That rubble will be removed by way of a tunnel that connects the bottommost relieving chamber with the gallery.

With relieving chambers in place the high chambers' roofing beams will be given needed assistance in bearing their own great weight.

Relieving chambers also allow thin wallstone to be used in the high chamber.

Once all the relieving chambers' rough beams are emplaced its limestone roofing blocks will climb the east face.

Emplaced at an angle, along each length, the limestone blocks would squeeze the topmost row of rough beams between them. This inward pressure would help the topmost row of rough beams support their own weight.

You may wonder at the great step pyramid.

You may wonder at its wide ramps and working lanes.

You may wonder at its very size.

Be reassured.

The womb that holds the inner pyramid, the outer step pyramid, is also the upper stone of that same inner pyramid.

The great pyramid will give birth to itself, like the legendary bird.

This is only fitting. It is a magical structure.

During the first phase of construction enough stone will be gathered, into a small number of mastabas, to construct an entire pyramid.

Then the quarry will no longer be needed.

The step pyramid will be the quarry.

During the second phase of construction, the structure will be transformed.

It will be transformed into one with many low mastabas. Mastabas low enough to be cut into a true pyramid.

Transformation will begin with stonecutters quarrying along the edge of each ramp and roadway to the end.

Then they will turn and quarry onto the next face.

They will remove a like width of that faces' working lane, up to the ramp.

They will quarry precut step pyramid stone. This will be the new corestone of the structure.

Stone that is removed will be sent up to build new mastabas, half the height of the former ones.

Stone will be going up all four faces.

Men will remove stone from the upper half of each ramp, roadway and working lane on a mastaba, one quarterwidth at a time.

If they remove a quarterwidth from the outside of each ramp and roadway, by turning and removing a like width of the working lane on the next face, they will create a new quarterwidth to the inside of each ramp; but only on the upper half of a mastaba.

Others will remove, lower, quarterwidths of vanished ramp and roadway. They will remove only the first two, lower, quarterwidths of vanished ramp and roadway.

One may already see a transformation taking place.

One may already see a mastaba becoming two.

The height of each mastaba will be half of what it was, formerly.

The width of the ramps and roadways, and of the working lanes, will also be halved.

This transformation will be the first.

With the final transformation of the great step pyramid, its once great mastabas will be low. They will be lower than the height of most men.

Did you not believe when you were told that this pyramid was created by magic?

The way each transformation will be done is said to have been thought of by a long ago magician, while in his bath.

He was looking at the surface of the bathwater and pondering mastabas. When he rose the level of the bathwater fell.

Some say he ran naked through the streets, shouting of his sudden inspiration.

I do not know if this story is as true as a creation story. I only tell it because it was once told to me.

Now; two new quarterwidths exist to the inside of each ramp, only on the upper half of each face of the mastaba.

Two outer quarterwidths are missing.

The lower halves of the two outer quarterwidths are also missing.

All missing stone was sent upward, to construct new tiers.

Third quarterwidths will now advance, turn at the end of the mastaba and form new ramps on a new, lower mastaba.

Stone that was their lower halves will remain in place.

On each face, the three conjoined sections of ramp will now march toward the end of the mastaba yielding stone, with which to make new height, as they go.

One will turn and join the section of ramp on the new mastaba, as its other half.

The other two will march straight into oblivion, leaving behind a new working lane.

The two quarterwidths of new ramp, on the new mastaba, will have left behind a new roadway.

Lower halves of ramp will remain in place.

They would now extend beyond a new, lower, mastaba were this not remedied. The transformation will then be complete.

During a transformation old ramps and working lanes will be used for a time, then new ramps and working lanes will be employed.

Explanation is provided so that the journey will be a coherent one. Measurements do not always have to be in given proportions.

I wonder if you have yet felt what I so often felt at Giza as I watched the construction unfold.

I looked beyond the labor and the stone and into the magic of it all.

I felt the thrill of being present at creation.

I sometimes thought I heard the legendary bird cry out once more.

Many would turn this into shells and colored stones.

They would turn all of life into shells and colored stones if one let them.

Now you begin to see just how a pyramid is created.

How was man created?

Did he acquire his passion for freedom after he lost his ancient instincts; or was it that passion that made him unable, any longer, to simply believe; that made him the prey of demons?

Do only some acquire a passion for freedom while others acquire a passion for belief, a need to be disciples of some school of learning, slaves to some greater truth?

The only companion I will have in my search for answers will be the demon. His great strength is that he does not lie. He always employs the truth.

I must turn this to my advantage. I must make this his great weakness. I must use this to make him my servant.

In this quest for truth, he will seek my destruction every step of the way. He prefers other goals.

And if I do find answers?

Then they will gather. They will suck the magic from all I discover. All I create will be turned into shells and colored stones.

Still, I must go on. What choice have we?

Much of what I have related, so far, about the great pyramid is what might have been. What was planned and may have been, if Suri and I had not gone to dinner at Nebris' house.

"Did you know your friend was a tomb robber?" our host asked me, toward the end of the meal.

I felt there was no danger. The way Nebri spoke, his striving for a conversational tone.

"I knew", I lied, making myself guilty of aiding Suri. "I knew it was never proven", I added, making a guess.

"Suri didn't linger for the inquiry", Nebi said to me.

"Did you?" he asked, turning toward Suri.

"What inquiry?" Suri asked, wide eyed. He didn't play innocent, he played the innocent.

Nebri smiled. "You'll both do", he said.

For what I wondered?

"Have either of you been to any large town lately?" he asked.

We told him we had not.

"The temples of the sun cult are flourishing and well guarded by the elite forces", Nebri said. "Many of those of other cults have been forced to close."

My old temple had not closed but had shrunk in size as the power of the sun cult grew.

Nebri went on, quietly talking.

Suri took another helping of dessert.

I sipped my wine.

Now you must learn a little more about our land before we continue the journey.

There are many lands and many gods. Wiser men know them all to be the same god, the god of creation.

Once our land was many lands, so we have many names for the one god. We also have more than one version of the story of creation.

In one, a sacred bird alights on a stone in the waters of chaos. With a single cry it ushers in coherence.

Many believe that same stone is in the possession of the sun cult. It is in the shape of a pyramid.

Our burial customs differ from those of other lands.

Here, many believe that as long as a mans' body remains whole and uncorrupted his ba, or soul, can roam this world as well as the next.

They believe such travelers still have power to influence events in this world. However, if they do, if they linger here and strive here, they must be fed. They must be nurtured with the ka within food, with its essence.

Our pharaoh is both our earthly leader and our spiritual one. A position of the gravest responsibility. Yet, in his lust for power, our pharaoh has laid claim to divinity.

If a man lays claim to divinity because he has great earthly power and others believe this claim, or pretend to, how much insult is done to heaven?

If his children make this same claim, how much further insult is done to what is holy and eternal?

It mocks heaven and heaven will not tolerate such mockery. Power is best left in heavens hands. Heaven knows best how to wield it.

Those who lust for power arouse an all prevading wrath. Wrath which springs from the very souls of men.

Heaven is the cause of that wrath. Heaven has imbued mens' souls with a sense of justice.

Heaven need not create things. Heaven but creates forces. Forces that go on creating.

Heaven is the cause of all causes.

Let those who love evil remember this.

I must here tell of an obscene practice that has sprung up among the rich and powerful. The priests that pander to them, the priests of the sun cult, now teach that man-made things have a ka and can, thereby, enter into the next world.

So fools now fill their tombs with grave goods, seeking to pollute heaven with their gold.

Greater evil is done.

Empty words are added to the sacred texts. Fools are told these empty words are charms. Charms that will insure their passage into heaven. There are those priests who will recite these words at a mans' funeral.

The creator of understanding is seen as lacking in it.

Once everyone was taught that they could only enter heaven by doing good. That good was even seen as being weighted by heaven. Now everyone is told what great good lies in petty selfishness.

Is it any wonder so many bewail a coming darkness, while others quietly await it?

Nebri also spoke of the rise of the sun cult. He told of their corrupting influence on the previous pharaoh. He too had been seduced into claiming divinity. He too had chosen to be entombed in a pyramid, the symbol of the sun cult.

The sun cult had gained power by pretended devotion to him as a god on earth. He had allowed the formation of an elite force, under the control of the sun cult. A force that would protect the interest of the royal family; for now.

After his death the sun cult used their influence with his son and heir, Khufu, to have Giza made the site of a great necropolis. Its many tombs will stand in the shadow of Khufus' pyramid.

Many tombs because the sun cult's influence was continually growing among the rich and powerful. With them it was a corrupting influence. It turned selfishness into a virtue, then preached it.

In some parts of our land, people were going hungry. Most of the grain was being exported. Those who owned the granaries paid the farmers little for their grain then resold it, at great profit, beyond our shores.

The need for elite forces, that could be used against our own people, was growing.

Many of the royal family, Nebri told us, were well aware of the growing wealth and power of the sun cult. They were also aware that their own wealth was shrinking, along with the esteem people once had for them.

"What could be done?" I asked. I now knew the dinner had an underlying purpose.

Nebri would not be rushed. "Did you know that the tomb of Khufus' mother was violated?" he asked, in response.

I shook my head.

"Her body was taken", he said.

"Was anything of value left behind?" Suri asked, much to my surprise.

"Gold", Nebri answered.

Suri nodded, then returned to spooning yet another helping of dessert into his bowl.

"Giza will be a gold mine for the sun cult", Nebri said.

Giza would be safe. That was being promised, as a certainty. It was common knowledge, anyone who strayed near to the newly built tomb area would be held and questioned by the elite forces.

No other place was safe for the bodies or grave goods of the rich. As Nebri had just told us, the tomb of Khufus' own mother had been violated.

Much wealth had already been invested in the Giza necropolis. Some, from the richest and most powerful families, had already been entombed there. Others had finalized agreements with the sun cult for their perpetual care, as had Khufu.

More tombs were building.

Giza would be a great many gold mines.

It had been decided, Nebri told us. Action must be taken.

For the good of the royal family, action must be taken.

For the good of the people, action must be taken.

"Khufu lives in terror of anything happening to his body after his death. Anything that might affect his afterlife", Nebri told us. "That fear could be used against the sun cult", he added, more thoughtfully.

I played my part.

"How?" I asked.

"Sink the plugstones", he replied. He paused to sip wine in the lengthening silence.

His was a clever idea. No pyramid could be sealed without plugstones. In the years it would take to finish the pyramid other plugstones could be made but to what purpose?

In this pyramid, once construction went beyond the inclined gallery no plugstone could enter it.

If it couldn't enter the inclined gallery, it couldn't be used to seal the pyramid.

The sinking of the plugstones would be the sinking of the sun cult's ambitions.

"Soon the plugstones will be sent to Giza aboard a single ship", Nebri said. "The captain of that ship happens to be a brother to one of the captains your team employs to bring deck cargo to Giza."

"We are to find out if he can be bribed?' I asked.

"He can be bribed", Nebri said. 'He is a devout believer in the god Khufu, so he is corrupt. You need only find out his price."

"He may need help to sink the ship", he added, "the twins are sailors."

By now I knew my meeting Nebri had been no accident. He seemed to know everything about us. We knew nothing of him. Things were moving too quickly.

"Give us time to talk to the others", I said.

We made arrangements to meet again.

Suri and I took our leave.

We walked a good way before Suri broke the silence.

"What was that powder on the dessert?" he asked me.

"Why" I asked, "did you like it?"

He laughed.

"Cinnamon", I told him.

"It was good", he said.

We both laughed.

We walked a good deal further before he said, "For the tomb of Khufus' mother to be robbed those in authority had to take a hand."

Further along he said.

"That's how it's done. We worked with the officials, with those who were supposed to guard the tombs. I don't know who. I was not that high. My share was small."

Still further and he said.

"Sometimes men are paid to rob a body. It may be that an enemy believes once the body is destroyed, the soul can do him no harm."

Later, he said.

"Some who rob tombs have family or friends who were sent to the mines. Often they take revenge on the bodies of those they see as the enemy."

I knew Suri well enough to know that mingled with his words was an effort to let me know that it may well have been Nebri who caused the disappearance of the body of Khufus' mother.

It wasn't much further along that same road that a young man who had fled the world, some time ago, realized he had now been restored to it.

All the team members followed the divine in a less fiery form, as did almost every worker on the pyramid. The sun, like Khufu, was far removed from us all. This would make my task easier.

The next day I spoke first with Nin. This was as it should be. He was the oldest and our team leader. I simply told him the story of the previous evening, then waited.

"I was a soldier", Nin said. "Some time ago my unit was sent to the western delta. We were part of a larger force. I was a sergeant."

"We were sent to remove civilians from land they claimed was their own. Some had lived there for generations. We were told they were Libyans, perhaps some were."

"There were those who felt we were taking too long. Men who wanted that land. Men who already owned a great deal of land."

"Other methods were demanded. The elite forces were sent for. These are men not very good at soldiering or following orders but good at murder and torture."

"Before we left there were words between some of them and some of my men, over the killing of a woman. Blows were struck."

"Our officers ran up and stopped us. We listened to our officers, that was lucky for them. Only one of them was killed."

"I was the sergeant. I was held responsible. I was forced to leave the army. I came to Giza."

Here, Nin paused.

"We will all be together in this", he continued. "I will talk to Seti. You talk to the others."

Would I were a real storyteller and not just a scribe. I would tell you a story of a land that was two lands, a black-land and a redland.

There were two people in this land that was two lands, those who were people of the blackland and those who were people of the redland.

There were two great emotions in this land that was two lands, empathy and its opposite, fear.

The people of the blackland knew empathy and re-spected altruism.

The people of the redland knew fear and respected power.

The god of the people of the blackland lived in the fruit-fulness of their soil.

The god of the people of the redland lived in the power of their sun.

Now comes a strange part of this story that is also a his-tory. The holiest men of this land that was two lands had always sprung from those of the blackland. Yet all of the most powerful priests were men of the redland. No one has ever shown why.

"My father and the father of Amus earn their profit the same way old Pooth does", Abi said, "by helping others to profit, not by stealing, like so many new men of business.

I will talk to Amus", he said to me. "He will be with us in this."

Which left only the twins uninformed. After I remedied this they surprised me, once again, by being themselves.

"This captain that is to sink the plugstones may need help. Remember, we are sailors", were the words that so surprised me.

"I never felt so rich", Suri said.

I had just passed him the clay jug of beer. I didn't re-spond. I was rapt, in stars.

I too felt rich.

Only a desert sky could supply such treasure, such a flow of wealth each night.

"A good deal can be said in favor of living dangerously", Suri said, "but there is also a lot to be said for life here, at Giza."

He drank from the jug.

"Being part of a good team keeps your work days from being just dull toil", he said. "It also does a lot for your rest days, thanks to Pooth."

He offered the jug. I shook my head. I'd had enough.

"Some of this will have to be poured upon the ground", he said. "Strange, not one of our team is much of a drinker. By myself I'm an oddity in Ragtown. What would you call a team of eight such oddities?" I half heard Suri ask.

"A bigger oddity?" I ventured, while stargazing.

He shook his head. Soon he rose, left the jug, and walked further out into the desert.

I watched the stars and pondered free will.

Without freedom how could we possibly sin?

Without freedom how could we possibly hear the demons' voice?

When I was at school, an explainer wrote a scroll on free will. In this work he imagined a hurled stone suddenly endowed with consciousness. He claimed that the stone would conclude the course it was traveling was its free choice.

Moreover, he said, the stone would be right.

He was, afterward, held in great awe.

Would every rock come to the same conclusion I wondered, after half a jug of beer?

Suppose, just suppose, some rocks are hesitant, less sure.

Suppose there are others that, given consciousness, behave a bit impulsively.

Freedom is the offspring of the marriage of imagination and doubt.

Freedom will not be imprisoned by reason.

Freedom will not be understood.
If it were, it would not be freedom.
Chaos is not subject to reason, neither is freedom.
Neither can be explained.
Which is a suitable definition for either.

Suri was right. Some beer had to be poured upon the ground. I did so and walked back to Ragtown, carrying the empty jug.

The Fifth Scroll

The largest single feat of internal construction will be the walls of white casing stone that will enclose each tier of the inner pyramid.

There will be over two hundred of them.

Each white stone will lie along the slant of the finished pyramid.

Each will be cut along that angle during the third and final phase of construction.

A length of casing stone will lie along a tier the same way it once lay in the ground. In this way each stone will fit, side to side, with little effort.

Tiers of casing stone will overlap, alternately, at the corners of the structure. This way they will be stable.

Now it is time to see how the inner pyramid and the step pyramid are joined together.

Now it is time to see how two structures, with incompatible tiers, become one.

The inner pyramid will stand atop a pedestal.

So too, will the step pyramid.

So too, will the empty area beyond the ramps of that step pyramid.

We shall lay aside the fact that the middle step pyramid will become corestone for the great pyramid. Nor will we discuss corestone.

The great step pyramid will have four sets of ramps, one at each corner.

They will be wide ramps but their inclines will only be one cubit deep.

With the final transformation every mastaba will be three times higher.

The south, the west, and the north ramps and roadways will be constructed. The roadways will be six cubits high.

A partial east roadway will be constructed. It too will be six cubits high.

It will not reach the east incline. It must fall short by thirty-six cubits, the distance needed to construct a ramp to reach the top of the east roadway.

An additional area will also be empty of stone. This area will be a jumping off place for stone moving into the inner area.

The first tier of inner pyramid stone will now be laid and will also fill the jumping off area.

Stone moves across the pedestal on greased planks. Pushsleds move the stone, levermen emplace it.

The first tier of east ramp will be laid down.

By way of the east ramp and the jumping off area a second tier of inner pyramid stone will be laid.

The casing stones of the second tier will be set back further than the casing stones of the first tier but the face of both tiers will be made even.

A second tier of stone will be laid atop the jumping off area and atop the east ramp.

By way of the new height of east ramp and jumping off area a third tier of inner pyramid stone will be laid.

The casing stones of the third tier will be set back further than the casing stones of the second tier.

The lower part of the third tiers' face will be even with the face of the two lower tiers. It will be even with them to a height of six cubits.

A layer of stone will also make the jumping off area and the east ramp six cubits high, as high as the east roadway.

The upper part of the third tier of inner pyramid stone will be set back. It will be a part of the next step of the inner pyramid.

It will be as though there were a flight of steps just beyond what will be the slant of the finished pyramid.

These steps are of excess stone.

The tierstone of the step pyramid will sit atop these steps.

A second level of south, north and west ramps and roadways, six cubits high, will be constructed.

The next segment of east ramp will be constructed atop the next thirty-six cubits of east roadway; after a certain quantity of six cubit high, step pyramid, stone is emplaced along the remaining length.

Another jumping off area at the top of this new length of east ramp, will be left vacant of six cubit high, step pyramid, stone.

As a new height of ramp and jumping off area is laid down a stone bridge, that will adjust its height to that of the rising inner pyramid, will also be laid down.

It will run between six cubit high, step pyramid stone

The inner pyramid will contract as a mastaba grows higher but the step pyramid will gain in width.

As a mastaba grows higher each new bridge will become longer.

Corestone, casing stone and special stone will travel over a bridge.

A mastaba of the high step pyramid will be eight tiers of step pyramid stone.

The tiers of both inner and outer pyramids will meet atop a finished mastaba.

The descending corridor will rise above ground level through tierstone; the tierstone of the middle and high pyramids and the tierstone of the jumping off areas of three step pyramids.

Explanation is provided so that your journey will be a coherent one. Measurements do not always have to be in given proportions.

Once, some time ago, there was debate between two groups of engineers as to how high a step pyramid should go before transformation began.

One group of engineers, skilled in math, argued for a lower height.

Another group of engineers, skilled in construction, argued for a higher structure, for more stone.

So a competition was arranged by Pharaoh.

Two step pyramids were constructed simultaneously.

The higher step pyramid was successfully transformed. It became the royal pyramid.

Final transformation of the lower step pyramid was thwarted by normal wastage.

It was decided to build another pyramid atop it, using its remaining stone. This upper pyramid would be given a shallower angle.

It gave the finished structure an odd look.

Suri remarked that a great deal could be said in favor of living dangerously. I was coming to understand just what he meant.

He had meant a great deal could be said for the excitement of living dangerously.

A great deal can be said for excitement.

There was new mood in the air.

During our meals the twins, when telling of the many dangerous places along the river where a ship might run against rocks, become almost verbose.

It could be made to happen in such a way that it would appear accidental and the crew would have time to abandon her.

We felt more alive.

We felt we were already striking back at those who were undermining our society.

We felt we were already striking back at those whose greed and lust were destroying our nation.

I shifted, trying to get comfortable on my chosen stone.

I wondered if what we call the ka was what others, in foreign lands, refer to as the conscience. They believe this conscience to be an inborn moral code.

This is instinct. Something I believe we lost, long ago. Something we traded for imagination and doubt.

Something we traded for freedom.

Yet these same people believe some are born without a conscience. Foreigners are difficult to understand.

They are unaware of the two people and the lands, those of the blackland and those of the redland.

The ka of a redlander makes him fear such a thing as pity. He believes others will use pity to gain power over him. So he hardens his heart, believing he is strengthening himself.

This is how redlanders are born to think. It is their ka. It is their nature.

Nebri was a darker shape against the background of night. He drew closer. I rose to greet him.

"Did all go well", he asked, after an exchange of greetings and as he took a seat on a nearby stone?

"Our captain does not foresee any problem with his brother", I said. "He wished an advance payment of gold to bring to him. I told him how payment would be made."

"It's strange", I told Nebri, "I think he half believes that Khufu is divine."

"Not so strange", Nebri said. "Many believe all that they are told by those in power, if it is repeated often enough."

"Few workers do", I said.

"They live in a different world than your captain", he said. "Do not worry. Your captains' love of gold is strong."

"Suri is annoyed with me", I said, conversationally. "He believes I don't yet realize our entire team was hand-picked. I realized this before I spoke with them about the plugstones."

"Anan is a good man", was all Nebri said.

"I'm surprised he didn't pick a team of head breakers", I said.

"He did", Nebri told me, "more than one. It's best to be prepared. Many things can happen."

"If it was you who had Anan promoted than you are closer to Khufus' entourage than I first imagined", I said.

Nebri slowly nodded.

"Close enough to know that even a pharaoh can believe a lie if it is repeated often enough. Khufu is no different than other men", Nebri said.

Khufu is no different than other men!

He is more powerful than all those, taken together, who govern our many nomes. He exercises influence over surrounding lands. He is feared by their rulers.

Khufu. 1 used the name without adding any title of respect as a small insult. Like many 1 opposed the power of the sun cult and the blasphemy it preached.

Nebri spoke the name in a way that sent chills up my spine. He spoke it as though he spoke of someone he knew. The way 1 would speak of Suri.

"The real enemy of our people are the priests of the sun cult", Nebri said.

"They lust after power. They preach power."

"They amass more and more wealth. They will control all the property set aside to maintain this vast city of the dead."

"I fear their power will grow so great that, one day, their high priest will make himself pharaoh."

Ambition, 1 thought, the opposite of hate.

As construction is the opposite of destruction.

But when that ambition is for more and more power, for hedgemony over more and more people?

They that rule must have the loyalty or the acquiescence of the ruled. These must see the evil being done to others as not being crimes because those others are not countrymen.

Then it will be easier to commit the same crimes at home, against countrymen, against those deemed disloyal.

The priests of the sun cult have ambition. Their god is but a force. They will guide this divine mindlessness. They will use spells and charms.

This will be for the enrichment of themselves and those friendly to them. It will be no different in the afterlife.

Wasn't Khufu already one with the divine force?

Wasn't his building of the worlds' greatest structure proof of this?

Only those loyal to Khufu and the sun cult will have an afterlife. The god Khufu will see to this.

Hate, 1 thought, the desire to destroy; the opposite of ambition, the desire to build.

1 would not allow myself the foolishness of hate; contempt, yes; loathing, yes.

More than ever I felt a determination to sink the plugstones to the deepest part of the river. Then let the priests of the sun cult cast their spells and work their charms. Let them try to raise the plugstones, while the worlds' greatest structure waited, unfinished.

"It is more than just the plugstones", Nebri said, as if hearing my thoughts.

"If our land is to be saved, the god of good and evil must be saved. If our people are not to be destroyed, the god of power must be destroyed."

Long ago heaven put us on the path to freedom.

Long ago we ascended above other creatures.

Heaven has granted us magical powers.

We can create.

We can see things as they truly are, not just as they appear to be. The semblance of justice is not justice.

We can see what is yet to come. Time is not simply an eternal present.

There was no god of good and evil. There was only a god of good.

There was an evil demon whose followers craved power.

When fed, that craving would grow. Those possessed would want more and more power.

We were dealing with a demon.

Just as doubt could be used to destroy a man, so too could power destroy. This is what demons do.

They would feed on human souls, as wild beasts would feed on human flesh.

However, there is faith.

We were given enough insight, enough understanding, to know that we cannot truly know.

Nor can we truly believe.

This is why we were given faith. It gives us the power to believe.

It gives us the power to know the divine. It gives us the power to pray.

It has even given some holy men the ability to hear the divine voice.

Some have faith in gold.

Some have faith in power.

Some have faith the dragon existed.

Some have faith the creator of that dragon exists.

"The priests of the sun cult view divine power as they view the endless and mindless rising and setting of the sun", Seti said.

It was some time after my conversation with Nebri. 1 was having one with Seti.

"Light a fire under a pot of water", Seti continued, "and when the pot is hot enough the water will boil. They believe this is what heaven is there to do."

"Khufu, once he reaches the next world, will instill more purpose in the divine", Seti finished.

"I believe in the god who gave us the sun", 1 said. "I also believe he gave us the colors of the rainbow and the scent of flowers, along with the sound of bird song."

"So do they", Seti said.

"Do they believe in truth?" 1 asked.

"What is truth?" Seti asked.

"What some search after", 1 answered. "What some value above all things."

"They believe in hammers and in pegs", Seti said. "They are the hammers and we are the pegs. That is their truth."

"Then there is only the god who boils our water?" 1 asked.

"There is only the god who created us all", Seti answered. "In our hearts we know this. Why else do we feel as we do?"

"Soon", 1 told him, "the twins will be making a journey. They will visit their fathers' tomb."

"Do many know the journeys' other purpose?" Seti asked.

"Only the team and Nebri", I answered, "perhaps Anan", I added.

"Men are slow to action", he said, "even against evil. "What is to occur will upset many. It is best no others know."

Nin said. "One day the men of many nomes will be arrayed against the sun cult."

"Khufu has granted some hereditary rulership of their nomes, in exchange for gold. These men are not about to tum their power over to the sun cult, as Khufu has done."

"One day I will leave Giza and go to live in one of these nomes. Nebri is right. One day Khufus' offspring will have lost their power. A priest of the sun cult will rule."

"Then my children will fight against and help destroy the sun cult."

It was simple and straightforward for Nin. He was a simple and straightforward soldier. I could almost hear the clash of his childrens' swords. They would fight well. There would be a family debt to be repaid.

Days later I waited in the same park where I first met Nebri. I watched our captain approach. He was now one of several captains that brought deck cargo to Pooth and our team.

In addition we had for some time supplied trustworthy workers to other enterprises in Giza.

When he reached where I waited I simply told him the words, the phrase, he would recite to Pooth. Words that would move Pooth to pay over a weight in gold.

Knowing Pooth, he would pay over the gold wordlessly. If the captain undid the canvas and weighted any gold in his hand Pooth would watch, just as silently.

After our captain turned and walked away I too turned and walked away. I joined the twins and the rest of the waiting team.

We would walk to the ship. If we took our time our captain would not be far behind us when we arrived.

Beneath their garments each twin wore a canvas belt that held gold. Others carried their seabags.

Last night Nebri had given the belts to me. I had passed them on to the twins. They held the advance payment to our captains' brother. Payment to sink a state ship. The ship carrying the pyramids' plugstones.

Late last night, upon seeing the gold, Nin had expressed concern for the twins' safety. So had others. Suri, my former mentor, felt he should join the twins on the voyage. I was forced to tell them all the story Nebri had told me.

Anan had earlier visited the ship with a gang of his headbreakers. Most men in our land, in spite of the heat, went clothed. Anan and his gang went to the ship naked.

Anan felt this was more intimidating, Nebri told me. Perhaps because their collective nakedness revealed more scars. The only item of clothing among them was Anans' expensive head scarf.

While Anan expressed his concern, to our captain, for the safety of his dear nephews, the headbreakers stared down members of the crew.

The crew members were at a disadvantage. They were men who wanted no trouble. The headbreakers were men who sought it.

I had then told Nebri a story.

Once, Tu told me, a man had slapped him while he and his brother were in town. I forgot just what had prompted this revelation.

The man had a friend alongside him. Both had been drinking. They were looking for someone who could be made to buy them another drink.

Tu had been shocked.

"Why" he asked the man" did you slap me?

For reply the stranger made to slap him again.

An angry Tu left both men lying in the gutter. One had a broken wrist and the second, who had attempted to assist the first, was curled up in a ball, moaning.

"Didn't you help Tu?" I had asked Ty.

He shook his head. "There was no need", he stated frankly, "there were only two of them."

Nebri laughed.

After Tus' story I never forgot that the twins had been raised on the river.

At the ship Abi admonished the twins to eat the wrapped sausage, that he had put into their seabags, themselves. It was too good to give away, he told them.

He was wasting his breath. The twins would gladly share all they had with any crew members who was kindly or friendly.

Both twins were blacklanders.

Late that night there was a meeting between Anan and myself.

He informed me that Nebri would no longer come to the worksite. He, Anan, would meet with Nebris' intermediary. Then he would meet with me to exchange information. It was time for all to be more cautious.

"Nebri tells me you believe we ascended from other creatures", Anan then said. He adopted a pose, his legs wide apart, his arms crossed.

"It was the divine will", was all I answered.

"He did this by giving us imagination?" Anan asked.

"Yes", I answered, without explanation.

"I'll grant he gave you more imagination than most", he said. "You also believe doubt is a demon?" he asked.

"Some hunt with a bow", I replied, "others use a sling."

"Some set snares", Anan said, "so be careful."

In the house of truth the mummies of animals and plants that lacked survivability are cataloged.

Animals and plants selected for extinction.

Bones and shells and leaves become mummies.

Mummies so ancient they are as hard as stone.

Some say great heat brought about the end of many of these things.

Some say great cold was the cause.

They call this natural selection

Some believe life becomes more specialized.

There are animals that live in the sea that once lived on land.

They are not fish.

They have lungs.

They breathe air.

There are those who believe the ancestors of these creatures needed only to live by the sea.

Random chance would do the rest.

They would gradually lose hair.

They would gradually lose their limbs and develop fins.

They would gradually develop appropriate tails, aquatic ones.

They call this evolution.

Some believe life becomes more diverse.

Others believe order can never come about by chance.

They say that randomness is infinite.

In some of our towns are gaming houses. Places built with copper and gold once held by those who believed that within random chance could be found a system.

Others say that even if there were an infinite number of such gaming houses, each would be profitable.

The owners of such places have faith in gold.

They know that those who have faith in random chance will bring them gold.

Men need faith in something.

Men need a god. Even if that god is only blind chance, men need a god.

It was a good many days later.

Suri and I sat on blankets beneath our awning while Abi and Amus played a game with tiles.

"Why not choose a simple shape," Suri said to me, "and seek to discover some new truth about it, instead of seeking to discover truths about mankind."

"Why should I do this?" I asked.

"The mind is an instrument for weighting other things", he replied earnestly. "It is no use trying to take its own measure."

"Are you also schooled?" Abi asked Suri. He meant no insult. He simply asked a question.

"Only in so far as I have taken opportunities to learn from others who are", Suri answered honestly.

"Just as questions and answers go together", I said to Suri, "so do doubt and creativity. Our creativity needs the darkness of doubt in which to shine. That darkness cries out for the light of our inventiveness."

"It is this combination of imagination and doubt", I said, "that gave us freedom."

"This is how we sustain individuality."

Doubt is disbelief.

One feels a supposition is false. Every supposition is part of ones' reality, a reality that is all supposition. One may then feel reality is false.

One cannot feel existence is false. Existence is not supposition.

I feel sadness. I also observe, I suppose, sadness in others.

Imagination is belief, at times shallow belief, as in daydreaming.

If deep belief and that deep belief is not part of ones' reality, then one may spend a lifetime making it part of that reality.

As imagination and doubt are part of ones' existence.

"Dhan is drunk", Suri said.

"We only drank apple juice", Abi said.

"Dhan can get drunk on water", Suri replied.

"As for your reservations", I told Suri, "some have maneuvered armies and rehearsed battles in their imagination."

"Astrologers have studied the heavens there. It is a vast workplace. Men have contemplated infinity there."

"If one man can go there to write a poem about mankind", I continued, "why cannot another go there to study mankind?"

"If one man has recreated creation there, why cannot another recreate mans' creation there?"

"With such knowledge I will, one day, conquer the demon", I said.

"Do not jest about demons", Abi said, without looking up from his tiles.

"I meant no jest", I said. 'I mean to use the magic heaven has given us to find out how we obtained our souls."

"Suri is right", Amus said to me, "no more apple juice for you." To Abi he said, while laying down a tile, "You lose again."

Leave Abi and Amus to their tiles for now.

Let me put aside, for now, the question of just how we obtained our souls, let it suffice that we have souls.

Who could look at the greatest of pyramids and deny this?

Who could look at this great work and not know who created its creators, those who arranged and rearranged its pieces, then sculpted them?

This structure will be the folly of one man, a man who would be a god. As such it will be a failure. Not one of its eternal houses will be slept in, not for a single night.

This structure will be the aspiration of many men, their prayer to a living god. As such it will succeed. Not one of its eternal houses will be lived in, not for a single day.

As a tomb for one man it would be monstrous.

As the prayer of many men its size and shape suffice.

Here would be a good place to end the first part of the journey. That part of the journey that is about how the great pyramid was planned.

Here would be a good place to begin the second part of the journey. That part of the journey that is about how the great pyramid was built.

The Sixth Scroll

It was weeks later than Anan said to me "you're a hard man to find."

"I'm usually here whenever I'm not in camp", I said. I meant the worksite.

Tonight Anan wore a tunic and his smile. A smile fed by his amusement at everyone else. He also carried a weapon; himself.

There's a wind from the south", he said. "Do you feel it?" he asked.

I felt no wind and told him so.

"My guess is there'll be a storm", he went on. "A bad one". He was opening and closing his fingers, clenching his hands, unaware.

Realization began to dawn on me. My heart skipped a beat.

"Oh, yes", Anan said, reading my facial expression as though it were a scroll. "The two little sailors did it. I never believed they could. Heaven works in mysterious ways."

The plugstones were sunk! I looked toward the river as the enormity of it struck me.

"Go back to camp", Anan said. "Let the others know, quietly. Tell them if they wish to go on living they best behave as everyone else does."

He turned and walked away.

I was grateful for his words. If he hadn't spoke, hadn't told me what to do, I may simply have sat down, under the weight of amazement.

Instead, I walked toward camp.

Above me, in the darkness, a darker shape moaned. It moaned and howled in pain and anger, like a beast.

Perhaps, I thought, it's only the moaning and howling of the wind.

But again, there was no wind.

The next morning Ragtown woke in consternation.

Those of its citizens that weren't religious were superstitious. So everyone knew, the sinking of the plugstones had been no accident.

The night before, I slept but little. I lay awake. I went over everything.

Nebri had arranged for the payment, by Pooth, through a third party. Pooth had no idea why our captain was receiving gold or that it involved the team.

But two members of the team were on the ship that sank with the plugstones. That would not be secret for long.

The team was not supposed to know our captain would do the twins the kindness of arranging a ride back to Giza for them, aboard his brothers' ship.

After a visit to their fathers' tomb, the twins would board the brothers' ship. It would be off loading cargo in that same town.

Next, it would go to pick up the plugstones. Then it would take the plugstones, and the twins, to Giza.

The twins would repay some of this kindness by standing watch. This was the story as it would unfold.

Had our captains' brother been cautious about dispersing the gold the twins had brought him?

I hoped so. There was nothing else I could do.

After the twins returned there would be a crowd around our awning, wanting to hear the story of the sinking; a series of crowds.

At the morning meal Nin and Seti were quiet, as they often were.

The usually talkative Amus was also quiet. I could understand. My own talent for acting was small.

Without Amus engaging him in conversation Abi too was quiet.

Suri exuded amusement.

He wondered, out loud, when the twins would return from their perilous journey.

He next wondered, again out loud, how Khufu had taken the bad news about the plugstones.

He guffawed.

Nin told him to put a lid on it.

"The twins are better sailors than many supposed", Nin said quietly. "They are also good soldiers", he added, even more quietly.

It may have been then that we all started to feel the suppressed excitement again. I know that's when my own resumed.

The plugstones were sunk!

This morning the whole camp knew.

Soon, the whole realm would know.

Let the priests of the sun cult try to explain it away.

As we left for the worksite I said to Suri "I imagine Khufu must have felt very mortal when he heard the bad news about the plugstones."

At the structure, work was well advanced on the second mastaba of the high step pyramid. The walls of the high chamber were to rise with this mastaba. There was no reason for work to have slowed.

The ridge will be leveled to the planned height.

The high chambers' floorstone will be emplaced.

Courses of high chamber wallstone will move along an ever higher and higher route.

This had been the plan.

We waited for the signal to begin work.

No foremen or general foremen were to be seen. Workers walked from group to group, exchanging rumors.

Many believed the structure would be made wider.

Which meant it would also be higher.

The capstone would fit.

Let another descending corridor be cut, to join the old one. The new descending corridor would be cut from inside the old one.

It would be in three stages; a downward stage, a horizontal stage and an upward stage.

After the funeral new plugstones would be available to seal the new entrance from ground level.

Some transformations would be performed on the present structure, making it less wide.

A new pyramid would then be constructed to surround this new, less wide, step pyramid.

A new step pyramid would surround the new pyramid.

When transformation was resumed, on the new structure, a new pyramid would emerge with a new entrance at ground level.

The old entrance would be covered over by the stone of the newer, larger structure.

Some said that too many knew where the old entrance, on the north face, was located. This would make it too easy for tomb robbers.

Very well, let a new tomb be made in the horizontal section of the new descending corridor.

Let a granite barrier gate be constructed to secure the southern section of that corridor.

Not all agreed this was what would be done.

What all did agree on was heaven was punishing the sinfulness of Khufu and the sun cult.

That day we worked half a day. There would be more half days. There would be short work weeks.

Work was slowed while new plans were made and implemented. This was just as well.

The temporary replacements for the twins were both strong and willing workers but each, as Nin said, had two left feet.

Now and then such a one found his way into the army, Nin told us, but never in pairs.

"This had been planned by someone with an evil sense of humor", he said.

The bottom of a lever has a wooden 'toe'.

It also has a 'heel' that pushes against the stone being moved.

Levermen 'walk' a stone by levering in cadence.

Laying their weight on a rope the ropemen heaved in unison, keeping pace with the pumping levers.

Our two temps lacked the ability to coordinate.

It was an astonishing lack.

It was also a painful one.

"They weren't just uncoordinated", Nin said, they were "anticoordinated."

A few days after the twin's departure we were all seated, aching and exhausted, under our awning after a very hard day of fighting the work.

The patient Seti, who would take any workers' part against other men, began slowly shaking his head.

"I don't believe it", he said softly. He spoke to himself, like someone who had too much to drink. "Just by chance, even accidently, they should lever properly for a time." His quietly amazed voice was only a whisper.

My body ached from fighting the work all day yet I found myself chuckling; a little at first, then a little more.

Soon, tears of laughter were running down my face. The rest of the team looked on as though I'd lost my mind.

A stone would move and all would pull with the movement, easing off when we felt the stone begin to slow.

Suddenly, the stone would give a jerk, because one temp levered so out of rhythm with the other.

We would be thrown off stride. Some of us would stumble.

After the third or fourth stumble Anan would yell "You're supposed to pull the damn stone, not dance with it!"

Anan now spent most of his time watching our team. We knew the watching filled him with a quiet glee.

Each stone jumped along in spasms. We were not about to complain. All knew who possessed the evil sense of humor that Nin spoke of.

Early on, once we realized how hopeless the temps were as levermen, Amus volunteered himself and Abi to work the levers. That was the day the rope took on a life of its own.

We would be pulling evenly, laying into the rope, I would be bringing back one foot when the rope would jerk, sending me and others into a stumble.

We would recover and try to pull in rhythm but to no avail. The rope moved in fits and starts.

"Heave", Nin would bellow.

Four of us would. The temps would pick up on the command late, then try to make up for their tardiness with zeal, leaving some of us sitting on the ramp.

"Get up off your asses!" Anan would yell. "The horn hasn't sounded yet."

Abi later told me he thought each stone was alive and had ideas of its own that day. Often, when he began to lift up on the lever, a stone would jerk forward.

The next time he levered, no one would pull.

Constant spasms were mixed in with the progress of each stone.

"Never, not since the world began, has working a lever taken so much concentration", Abi said.

It was late that afternoon when Amus voiced the idea to use the levers to beat the temps to death.

Abi would have laughed but he was too tired.

Amus went on, even more earnestly, they could claim the temps tried to attack them he cajoled.

When Abi realized he wasn't sure whether Amus was joking or not, it frightened him.

The twins finally returned.

The temps left, unmolested, and we heard the story of the sinking.

It seems that as the twins were coming on deck in the darkness, to take an early morning watch, the ship collided with a submerged wreck.

It was the wreck of a ship carrying flax. It had foundered on rocks earlier that week.

During the night, the current had freed it from the rocks and taken it into the path of history.

After the collision everything happened quickly.

The captain came on deck and took the rudder himself.

Once the ship was free of the wreck he tried to steer her inshore. However, she was taking on water rapidly. She was abandoned.

The ship carried small wooden cylinders along her bulwarks. They were sealed with resin. They would keep a man afloat indefinitely.

Each man grabbed one to carry overboard.

No one was lost, all survived.

The manner in which the plugstones were lost convinced almost everyone that it was the hand of heaven. It also convinced most of the team, especially the twins.

A few crowds did gather around our awning to hear the twins' story. Not as many as I had foreseen. As I have already mentioned, the twins were not storytellers.

The high chamber will be moved.

It will be moved to the south.

It will no longer be centered in the structure.

The area it will be moved to is occupied by elevator compartments. Double com-partments, that hold the thinner granite beams.

Some of these compartments will be sealed.

Beams will be buried within them.

The beams are useful only in sets of four. So an entire set, not part of a set, must be buried.

Other beams will be fabricated to replace them. They will be made of limestone. The softer limestone is unsuitable for eternity but will have to do.

The high chamber is being shifted so that an antechamber can be constructed.

This antechamber will compensate for some of the security that was lost with the plugstones.

It will sit between the south wall of the gallery and the newly located high chamber.

The antechamber will be narrow but long enough to hold four barrier gates, each made of granite.

Each gate will be one cubit thick.

A short passageway will connect the antechamber to the gallery.

Another short passageway will connect the antechamber to the high chamber.

Both passageways will be made of granite.

The antechamber must also be made of granite, so tomb robbers cannot tunnel around the gates

Where will all these shaped granite pieces be found?

Stone will be removed from the top of the high chambers' walls. The high chamber will be lower than planned.

Slabs of granite, intended for use as ashlars on temple walls, temples within the pyramid complex, will now serve as sections of barrier gate.

Work will stop on the downward leveling of the ridge. There will no need to make the ridge any lower.

The leveled area atop the ridge and the great step within the gallery will be one cubit higher than planned.

Not all the plugstones were aboard the ship that sank. A few, a very few, unfinished stones remained behind at the granite quarry.

They will be finished and sent to Giza.

They will be far too few to fulfill the plan for the ascending corridor.

They will be used only to plug the lower end of that corridor.

They will serve as a hindrance, not as a barrier.

Now the entrance to the ascending corridor, in the ceiling of the descending one, must also be concealed.

Concealment will play a larger role in the new pyramid. It is hoped that after the pyramid itself is sealed and enough years elapse, internal concealment will be forgotten.

It is hoped future generations will believe the structure was abandoned to any use other than its religious one.

There are other changes.

The entrance into the high chamber will still be concealed but the great step will not be removed.

It is felt that if tombrobbers did penetrate as high as the gallery they would know the plugstones had to enter it by some route.

If the step remained they may conclude, wrongly, that no tomb lay beyond it. As that tomb would block the sure route. The hindrance such a step would offer to a funeral may help reinforce such a conclusion.

They may then search for the, unfinished, middle chamber.

Now the existence of an underlying ridge may be guessed. That is of no importance to those now making the decisions.

The architect who planned the structure and who allowed almost all the plugstones to be sent to Giza aboard a single ship will no longer be playing a role in decisions concerning the structure.

Some of this I learned in late night meetings with a new general foreman. A general foreman named Anan.

It is one of these late night meetings that I now invite you to.

I was, Anan told me, to find out all I could about the sarcophagus now lying on the floorstones of what was to become the high chamber.

As a supervisor Anan could not go poking about other areas. It would cause concern.

I could often be found where I didn't work. Not only because of my interest in the construction but because our team needed labor for our thriving business.

Almost everyone had come to accept my presence anywhere. Poking around the high chamber however, might present a problem.

"There is one more thing", Anan said. "The man you call Nebri is really named Hemiunu."

"Hemiunu is the man who planned this pyramid", I said.

Anan nodded.

"He is first cousin to Khufu", I said. I struggled to keep amazement from my voice.

Again Anan nodded.

"He is the vizier of our nation", I said. It seemed too much to believe.

"You mean to tell me the vizier reached down to make you a foreman?" I said. And had Suri and me to dinner I thought, in continued amazement.

"Churchmen of the sun cult made me a foreman", Anan said. "Hemiunu enlisted me, quietly, before they did."

Why churchmen?

Because for the rest of the journey I choose to refer to them as such.

Imagination and doubt, a blessing and a curse from heaven. Once I believed I knew which of these was which.

I was younger then, when I so believed.

Many live happily with doubt.

Not just the perverse who live gaily in a land without meaning.

Not just those who seek to sow doubt or to create nightmares.

But those who believe only in what they see, in appearances. Those who see reality but dimly.

One needs imagination to see reality. Concepts are part of reality. All our values, all we value are concepts.

Beasts have instincts.

Beasts have no need of concepts.

Men do. Unless they wish to live as liberated beasts, some new form of monster.

Those who become churchmen, not priests.

Those who seek to become rich, not men of business.

Those of the royal entourage who seek position, not service.

All who assuage their fears with dreams of power.

There are only the dreams. There is no power over others. The almighty has seen to this. The almighty has given us freedom.

Men can be coerced or men can be led. They can be coerced with lies and torture or led with truth.

There will be a desert land our journey must traverse, the country of the redlanders.

A ropeman may look upon the structure as being built pull by pull but a pharaoh looks upon it as being built tier by tier.

By this time Khufu resided at Giza. He came not long after the plugstones were lost. His presence was testimony to his discontent.

His presence was the cause of friction. Some foremen felt a need to increase their abusiveness toward workers.

One day blows were struck. Foremen rushed to the scene to aid other foremen, as they must.

Preparations had been made by some workers. A cry went up. Workers also rushed to the scene.

Levermen gripped their levers in a different way, ready to use them as weapons. Some workers carried stones they had picked up along the way.

An angry foreman demanded to know why one worker carried a stone. The worker, just as angry, told him he could easily find out why.

Further from the scene of strife, work ceased. It ceased in a spreading ripple. Teams began to ease toward one another, forming loose gangs.

Cooler heads prevailed.

General foremen took a part. The fire of rage subsided. Seti and workers like him added their breath, helping to extinguish any remaining flames.

The confrontation came at a bad time. Skilled workers, who could easily find work elsewhere, were leaving the project.

Even unskilled workers were leaving, if they knew where other work could be found.

It began with the sinking of the plugstones.

The loss of Hemiunu left many feeling that, at best, the second best were now in charge.

Then came the shattering of a section of barrier gate. One of the former ashlars fell and shattered while being off loaded at Giza.

Now only three barrier gates can be completed and only three will bar entry into the high chamber.

So many workers drifting away from the worksite left others feeling that heaven had also turned away.

The religious felt it was an evil place.

The superstitious felt it was a bad luck site.

Had heaven turned away from the project or from Khufu and the sun cult?

When asked, all agreed. Heaven had turned away from Khufu and the sun cult.

Once all religious cults had been equal. The many names of god did not blind men to seeing the one god. A pyramid was seen as the symbol of this god, as the giver of the sun.

It was seen as holy by many and respected by all. To be entombed within a pyramid would be, to many, blasphemous.

Churchmen of the sun cult had come to Giza with Khufu. A rumor made its appearance with their own.

Khufus' father had sought the approval of the sun god, through the sun cult, before entombment in a pyramid, the rumor said. Ceremonies had been performed to gain the sun gods' approval.

Khufu had not sought the sun gods' approval.

No ceremonies had been performed.

One day Khufu is the embodiment of the sun god. The next, he needs that gods' permission to do a thing. Such is Khufus' divinity.

Setis' father worked near the high chamber. As a stone-cutter he was curious about what could now be done with granite. He had gotten more than just a look at the sarcophagus in the high chamber.

Advances in working hard stone were being made rapidly. There would even be a basalt court in the pyramid complex. Basalt is harder than granite.

By talking to him I found out the sarcophagus had cuts, badly made, inflicted by a jeweled saw that had gone awry. It had done so more than once. Finally, a proper cut had been made.

These defects were not the only manmade ones.

I heard of a second sarcophagus. It was kept in a secured building, where things of value were stored.

Rumor said it was made in two sections. The bottom section being lower than normal. The upper section being deep, more than just a lid. The whole being very well made.

I hadn't yet gained that one dependable description of it.

What information I had I sent to Hemiunu. I drew no conclusions.

Conclusions could be drawn. It seemed Khufu could be making preparations to be buried elsewhere, if things continued to go badly.

Were Khufu interred elsewhere there would be outrage from many powerful families. Families that had invested in perpetual care. Those that had already built and those planning tombs at Giza.

The heads of some of these families governed other nomes.

Were Khufu to be interred elsewhere the sun cult would lose countless fortunes. Khufus' divinity is to be passed on, to his heir. There will be other pyramids at Giza, many more tombs.

Were Khufu interred elsewhere a mock funeral must be held at Giza. A mock corpse must be interred in the inferior sarcophagus. Those involved, ruling family and sun cult, would surely keep the secret.

If things changed for the better Khufu was also prepared, with the second sarcophagus.

If rumor was accurate the second sarcophagus could be sent up the ascending corridor and into the high chamber, a section at a time.

The larger, poorly made, one could be broken up and its pieces removed.

Sometimes we wondered if the sinking of the plugstones would ever lead to the wreck of the evil plans being made along with those for the new construction.

Those who would serve the almighty must ever struggle against those who lust for power.

The Seventh Scroll

The innermost construction is stunted.

The high chambers' walls are lower than planned.

The tiers of the inner pyramid must be rearranged.

The tier that will support the first of the rough granite beams, as they travel to their final resting place, will be needed sooner.

Mastabas must not change.

Transformation will still occur proportionately.

We leave these headaches to those whose tasks are such things.

We continue the journey.

The limestone replacement beams, for the relieving chambers, will ascend the third mastaba.

The third mastaba of the great step pyramid will only be half the height of the previous two but its ramps and road-ways, its working lanes, will be fully as wide.

Before transformation begins the width of the ramps and roadways, of the third mastaba, will simply be reduced by half; the ramps shortened.

The relieving chambers' limestone roofing blocks will be emplaced atop the third mastaba.

These blocks will be shaped much like the ceiling blocks of the middle chamber.

They will be rectangular, with one lower corner cut away at an angle.

A block will be uncut when traveling to emplacement.

The bottom of a block will be its skid.

Wooden posts will be inserted into holes on top of each of the highest rough granite beams.

Four separate sections of scaffolding will be erected down the row of beams.

Each section will serve one pair of, shaped, limestone roofing blocks.

Members of a pair will face each other.

They will press against the scaffolding.

They will angle over the topmost granite beams.

Each pair of blocks will then be fractured, along that angle.

The higher part of each, separated, section of blocks will rest against the scaffolding. There will be little strain on that scaffolding.

Before the mastaba is completed each section of scaffolding will be collapsed, then removed.

When a section of scaffolding is collapsed, a pair of limestone roofing blocks will come to rest against the highest rough granite beams.

The squeezing will begin.

As the structure rises a corbeled arch will be constructed over the limestone roofing blocks.

Once the relieving chambers' limestone roofing blocks are emplaced all temporary supporting blocks will be removed from the relieving chambers.

In the gallery, high up on the south end of the east wall, will be a tunnel that connects the gallery and the bottommost relieving chamber.

Rubble will be removed, from the relieving chambers, by way of this tunnel.

Fillstone will be removed from the high chamber.

No two people share the same reality.

Many believe they observe true reality.

They project their own assumptions into that observation. They assume objects behave coherently.

This coherence, that they espouse, they call cause and effect.

It leads one to logic and to math.

Logic is an attempt to prove that reason is more than supposition. It attempts this by using, contrived, cause and effect examples called syllogisms.

Math is equations. Four is equal to four.

Curiosity follows reason. Curiosity is a harmless form of questioning, different from doubt.

It is also based on the assumption that coherency exists, externally.

Resulting suppositions are called theories. They are put into catagories called alchemy, astrology and so forth.

But theories are created; created out of whole cloth, by the imagination. Because this is so a day will come when theories, in direct contradiction to each other, will be.

More amusing is that some will continue to believe reason to be more than a larger supposition, even though contradiction will plainly exist in that world of supposition so loved by men of reason.

Doubt is disbelief. It is a deeper form of questioning than curiosity.

Human behavior can go askew due to doubt. Suicide can result.

The curious avoid doubt.

Yet the curious question human behavior. Perhaps because their questions are formulated to fit pre-existing answers.

When I was at school there were explainers who expressed ideas not beholden to reason.

Indeed, they questioned reason.

They questioned cause and effect.

Their thought gave rise to art and to geometry.

Art is creation that reflects existence. The beauty of art is beyond any beauty claimed for logic.

Geometry is a field for the inventive. Its formulas go beyond the equations of math.

For art and for geometry imagination is needed.
Imagination creates.
It conceived reason, which is but a larger theory.
Reason could not possibly invent imagination.
Reason could not possibly invent. Reason propounds.
Dogs bark. Cats meow.
The I am doubts.
The I am doubts even the explainers. This is why the works of all explainers will be overthrown.
Man, the creature of theories, will not be encompassed by one. Imagination will not be encompassed.
Doubt is now an instinct, part of our being.
We believe in nothing.
This is why we need faith.
Some have faith in reason.
Some, those with great conceit, have faith only in themselves.
Some have faith in god.
There are the things of existence and the suppositions of reality.
The I am exists.
Doubt exists.
Imagination exists.
Freedom exists.
Faith exists.
If you sometimes doubt the existence of reality, remember, that is what you were created to do.
Men create new realities.
Men create.
One cannot doubt existence.
One cannot doubt ones' own existence.
One cannot help suspect in gods'; because we suspect others exist; because we possess imagination.
Eyes can only see so far, even the eye of the mind.
You may find it best, sometimes, to search for truth where it was buried. To search for truth within yourself, where poetry and pyramids are born.

The I am believes in existence, not in those suppositions called reality.

No two people share the same reality.

At the worksite roofing blocks were emplaced atop the relieving chambers. The stone of a third, lower, mastaba lay all around them.

Khufu was unhappy with the pace of the work. His unhappiness trickled down to some of those managing the work but caused no strife.

A truce existed between them and those doing the work. To enforce it some of the worst foremen had been transferred to the quarry.

Such was the state of things.

We prospered.

We had more captains bringing deck cargo to Giza. More businesses sought the rest day labor of the pyramid workers we supplied.

Our captain and his brother had purchased a small ship with the gold Hemiunu, as Nebri, had paid them. They now delivered cargo by the shipload to the partnership of Pooth and our team.

Which now brings our journey to a room Pooth used for business. To an evening when our entire team was present, after a day of seeing to the unloading and warehousing of cargo.

Pooth sat behind his desk. The two chairs that normally faced it had been moved.

They were now to either side of it. Nin sat in one, Seti in the other.

I sat with Suri on a wooden bench that ran along a table.

The tables' opposite bench held Tu and Ty.

The table and benches Pooth used in his business.

Abi and Amus sat on stools. One at either end of the table.

Pooths' two daughters had brought the stools from the family quarters.

The oldest daughter, smiling, had brought one for Abi. The younger daughter, more shyly, had brought the other for Amus.

The two young women then carried in pitchers of beer. Food was being cooked. It would soon follow.

"The temples of the sun cult are to teach the mysteries of business to the sons of the rich", Pooth said. "There they will learn how to collude to fix prices and end competition", he added, without humor.

"And how to cheat their workers", Seti said.

"There may be no need", Pooth replied. "The next pyramid may be built by foreign labor, from the puppet states."

"There are laws against underpaying workers", Seti said.

"Who would dare call such labor underpaid?" Pooth asked. "They shall be called guest workers. They will be compliant. If they become loud and unruly they will be shown the door. None will dare mention they are cheap labor."

"Our people will not stand for this", Seti said.

"It is rumored there is to be a loyalty act", Suri said. "Measures will be taken against any deemed disloyal."

"Ah?" Pooth remarked, "a way to deal with the native born loud and unruly."

"It is said there will be no hearings for any arrested by the elite forces", Suri added.

"Why should there be?" Pooth said. "Why should there even be trials for these worst of criminals?" he asked mockingly, looking toward Seti.

Seti would not be baited.

Nin had started shaking his head as soon as Suri had mentioned the elite forces.

"The rulers of many foreign lands are hated by their own people", Nin said. "Their people know them to be puppets of pharaoh."

"Such a ruler questions; by torture, any he suspects of trying to displace him and throw off pharaohs' yoke."

"He believes torture makes his people fear him. A suspect's family, even his children, are not exempt."

"Such torture also serves to demonstrate, to pharaoh, the puppet's loyalty", Nin continued.

"Our elite forces aid in this. It is what these jackals were created to do, to torture and to teach other animals how to torture."

"What about those who employ them?" Abi asked, quietly.

"They are just as evil", Nin said, angrily.

"More." Abi said, just as quietly.

Now it was Pooths' turn to shake his head.

"Madness!" he said. "Any land would offer what it had, in trade, for the copper and gold that we mine."

"Who else has papyrus to trade, or linen in such quantity?"

"No land exports as much grain as we do."

"Trade would flourish even if foreign rulers were not made puppets", Pooth finished.

"Dhan" Amus asked me, "why so quiet?"

"I was thinking about the ka", I said.

"No man is an atheist. All have a god. All have faith.'

"I have faith that truth exists", I said. "Setis' truth is justice."

"We both believe the divine exists, we just have different gods."

We see with the eye of the mind.

We see with the eye of the soul.

"While I was at school a teacher, who was dead to god, wrote a work and then went mad."

"In that work he declared that god was dead."

"He claimed the new belief, that man was but another animal, had destroyed god."

"He and others, blind with great conceit, took the natural second step. They sought to take gods' place among men. They wished to be seen as higher beings. Their reverence is for power."

"For them god is dead", I continued. "A demon has replaced him. The demon of power."

"For others, wealth is power. Their reverence is for gold."

"For another, mighty armies make him almighty."

"For yet another, position brings power. He struts about, swollen with his own importance."

"The lowest become elite forces. Devoid of empathy there is only the aloneness of self. The screams of others fill that emptiness with a feeling of power."

'Each needs to be an object of awe or fear to others. Power is their god."

That conversation did not end there. It was to resume again with Anan.

Hemiunu remained in touch but now my meetings were with Anan.

Many of these meetings were about the teams rest days.

We now worked as relief men. This way we did not all work the same days each week. Our schedules could be more flexible.

Our business was growing.

I must now, once more, invite you to one of those meetings.

Anan paced. He began to speak as soon as I sat.

He told me he was not unlike Seti, yet different.

He hated bullies, so in that sense, he too hated injustice.

As a young man he had killed a bully, from a family of bullies.

During a fight with Anan the man's eyes had scanned the ground, looking for a rock.

So Anan grabbed one first. He beat his opponent to death with it.

'It was that day", he told me, "I discovered that justice is best exacted with a rock."

He left home.

His nome was one where feuds were not uncommon. He left word for the dead man's family.

He told them what would happen to certain of their youngest children if his elderly father were harmed in any way and he must, secretly, return.

"You would have killed the children to prevent further injustice?" I asked, mockingly.

'Exactly", he replied. "To you, justice is a concept. To me, justice is something one exacts; after the sun has gone down."

He then spoke of a rocky, hilly area to the south, an area with many caves. He described the way to one cave in particular.

"Here lives a hermit many call holy", he said. "You may wish to visit him one day. He is much like you. He can't find his ass with both hands either."

Beyond this area of caves, further south along the river, leading churchmen of the sun cult dreamed of one day erecting a temple.

He knew because he had formed a relationship with one of these leaders, a talkative one.

This temple would unite the old pyramid area, to the south, with the new pyramid area, here at Giza.

This unholy temple would be used for human sacrifice.

It would be the first of many.

Human sacrifice would arouse peoples' fear and increase their awe of pharaoh.

It would also arouse their fear and increase their awe of the sun cult.

Each has a name.

Each lives in their own reality, a reality of suppositions.

Many of these realities have suppositions in common.

They are not shared realities, though many may think so.

Each lives in another reality, a different one.

It is a reality of imagination.

It is a reality of hopes and plans, of dreams and fantasy.

Most often, it is a reality unknown to others. As, once, the elevators of the great pyramid were known but to one man.

Now and then one of these realities, of imagination, washes away realities of supposition.

A new era is born. Power changes hands.

Often, this is why the new reality was conceived.

"There is a split in the royal family", Anan continued. That same split existed in society he pointed out.

Those who saw the danger of the sun cult and those who were part of that danger.

"The night is fast coming when we must all pick up rocks or find ourselves worshipping at unholy altars", he warned.

Which world I wondered, a world of unholy evil or Anans?

Which was when the more solid world began to tremble beneath us. Anan did a dance, to keep his feet. I debated whether to try to stand or to ride my moving stone.

The quaking grew more violent.

Suddenly, from the structure, came thunderous cracking sounds, one hard upon the other, some together. For brief moments, the earth thundered at the sky.

The sharpness of the sounds was dulled by the overlay of limestone but I recognized them for what they were. The massive granite beams were breaking.

The quaking suddenly stopped. Just as abruptly, Anan vanished into the darkness. I knew why. Anyone near the structure would be held suspect by the superstitious.

I followed Anan.

Inside the gallery plugstones swayed, timbers creaked.

It would be some time before all the damage was assessed. That assessment we received from Hemiunu, through Anan.

The antechamber was unscathed. This was because, Hemiunu believed, it sat atop the solid stone of the ridge.

Beams were cracked and displaced in the relieving chambers.

The high chamber had been badly shaken. Stone was displaced in every wall.

The greatest damage lie along the top of the south wall. Here, every one of the great granite roofing beams had broken.

Hemiunu advanced a reason for this.

To the south of the high chamber were long elevator compartments. Each held two of the thinner granite beams.

When beams were abandoned sand was removed from compartments until the abandoned beams lay just below the surface.

Mortar was then added, as fill.

Part of the width of the high chambers' south wall now rested on one of these abandoned compartments.

With the quaking the already compacted sand, in this compartment, was further compacted. The abandoned beams sank. The fill sank. The result was a slight depression.

The south wall tilted into that depression. All the granite roofing blocks broke, along a now higher edge of wallstone.

Structural stone also reacted to the quaking and the depression. It shifted toward the depression and against the south wall.

Disaster was averted.

Hemiunu felt all this movement took place rapidly.

The great roofing beams remained in place. Held by the weight of granite lying on their unbroken ends.

Nin felt life would be safer in one of the southern nomes.

"There, people have less love of power", he said.

"What of their reputation for feuding", I asked. "why are feuds fought?"

"Imagine", I said, "a churchman of the sun cult one day telling such people that peace can be purchased between two warring clans by an individual from each offering themselves on an altar of the sun cult."

Later Seti, said, "The only churchman of the sun cult that I ever knew felt more for beasts than for people."

"He viewed men as no different than beasts. Any difference that he saw was one of degree."

"He saw aspects of the divine in beasts", Seti continued. 'He had statues and figurines with human bodies and animal heads. They represented his vision of divinity."

Those who believe men no better than beasts become no better than beasts.

Such is the power of the imagination
They see a god with beastial qualities.
They would sacrifice others to it.
Killing excites them.
Being craves excitation.

"In northern nomes", Seti continued, "people such as him are bringing back the crocodile. They preach that all life is sacred. They mean that no life is sacred."

"Hemiunu is right", I said. "A great many will believe any foolishness, if it is told to them often enough."

One of the twins asked, of redlanders, "Why do they hate so?"

Self love is the only love redlanders know. Self interest is their chief interest. Often, it is their only interest.

Fear is the chief emotion redlanders know. It is fear that drives redlanders to seek power and dominion.

If the royal family was beset and split into factions, if pharaoh himself had been seduced, what chance did our people have against the coming evil?

Doubt can destroy. Men can destroy by employing doubt. A little doubt can go a long way. It is like the venom of an asp.

This is how new realities are created. Men doubt the old realities, often, because they are old.

Men are not like lesser creatures. Men will not graze the same ground their fathers grazed.

Realities change. Eras come and go.

After the earthquake Khufu took to his bed. He seldom rose and rarely left his room.

Thoughts of what a future earthquake might do to his future existence, in the damaged tomb, may have been the cause of this.

One of his sons was against the sun cult and any further increase in its power. After the earthquake many of the most influential in the land gave this son their support.

The power of the sun cult was checked, by the power of an earthquake.

Is thought of a mountain heavier than thought of an ant hill?

Is thought of a red sunset truly red?

Is thought of grass truly green?

How heavy is thought?

What color is it?

You can feel the air on your skin when it moves but thought seems weightless.

This less airy thing than air, is it the poetry or are we?

Who has not heard the old saw, "I think, therefore I am."

The I am is not thought.

The I am conceived the thought, then concocted the slogan.

One may think of an antelope and one may think of a unicorn. The antelope is real, I suppose, the unicorn not.

What is real and what is unreal cannot be made equivalent, even in thought.

One could feel a sense of security and go to sleep soundly thinking of a few pieces of copper, which one needs greatly, tucked safely away in ones' room.

One could feel desperation and lie awake, thinking of that same amount of copper, which one needs greatly, but does not have.

Thought about gold one has, may be in the nature of planning or may be in the nature of gloating.

Thought about gold one does not have may be in the nature of planning or may be in the nature of wishful thinking.

It could also be gloating, if one were mad.

Thought of a unicorn would not be thought of an unreal animal. There is no unreal world, filled with unreal objects.

It would be thought of the unreal only.

We think the unreal. We sometimes make that unreal, real.

We think the real. We sometimes make that real, unreal.

The moonscape of the lunatic, the stage of the fantasizer, the visual frame of the artist, are all different.

Delusion and fantasy and planning are not the same.

None, however, can always tell the difference.

What we feel to be true, using imagination, we will often strive to make so.

We reach down, inside ourself, and create. We call this inspiration.

Wherein dwells that invisible self; that self that supplies that inspiration?

Is that self in its knowing, amused by our lack of it?

How else do our mistakes become laughter, our sins become tears, our deaths become eternal life?

Fire, some say, is a gift from heaven.

So is that fire of inspiration that lights a deeper darkness.

Powers were at work deep within the earth creating, besides earthquakes, jewels and gold mines.

Other powers created storms and forests and grasslands.

We create groves and wheat fields, families and villages, nations and gods.

It is what we were created to do.

We are beings large enough to build pyramids.

The Eighth Scroll

The quarry is a growing, three dimensional structure. It must be shaped. Roads and ramps must be planned.

The quarry crew will shape the quarry.

The ground crew will take away the rubble, while the pyramid crew is moving the largest and heaviest of stones.

The quarry crew will level ground for each pyramid.

They will create roads.

Stone that is quarried will be large stone. If it were not large stone what is quarried away would be greater than what is quarried.

Men working different sections of the quarry will compete.

They will not see it as a trick to get them to work harder.

Men compete, without tricks.

Bonuses will be paid. Stone will be quarried on schedule.

At the center of the cleared area will be the bedrock pyramid. Here, leveling will be upward as well as inward.

The bedrock pyramid will be surrounded by the low pyramid.

Construction of the low pyramid will remain close to the shaping of the bedrock pyramid. Rubble must be disposed of, material must be delivered.

The low pyramid will be surrounded by the low step pyramid and its bottommost cleared area.

When the decision is made to construct the middle pyramid more ground will be leveled.

The low step pyramid will be surrounded by the middle pyramid.

The middle pyramid will be surrounded by the middle step pyramid and its bottommost cleared area.

The middle step pyramids' ramps and working lanes will be wide enough to accommodate the middle chambers', unformed, peaked fillstone and its, unformed, roofing blocks.

This stone will be deployed atop the second mastaba of the low step pyramid, when the ramps of the middle step pyramid reach that height.

After stonemasons complete their work, each roofing block will be slid into place along peaked fillstone and atop wet mortar. When that mortar dries each block will become one with the wallstone beneath it.

When the decision is made to construct the high pyramid, more ground will be leveled.

The high pyramid will surround the middle step pyramid.

The high pyramid will be surrounded by its own step pyramid and that step pyramids' bottommost cleared area.

The rough granite beams, for the relieving chambers, will travel a wide road that stretches from the river to the center of the high step pyramids' north face.

They will be deployed along the high step pyramids' north ramps.

When that structure reaches the proper height they will be deployed atop the second mastaba of the middle step pyramid, where preparations will have been made for them.

The high chambers' roofing beams will follow after the rough granite beams.

It had been planned to deploy them atop the second mastaba of the high step pyramid.

During construction of the low pyramid, the pyramid crew will be small compared to the ground crew.

During construction of the high pyramid, the ground crew will be small compared to the pyramid crew.

As construction expands there will be more places on a working tier where stone is being emplaced.

There is only the all.

Our great river is part of the all.

Many call the creator the first cause, uncaused.

I am not concerned with supposition.

Suppositions are a poor reflection of the all.

Why is it those who put their faith in the perceptiveness of reason often cannot see?

Hope is not supposition.

Hope is part of our existence.

Hope tells us things can change, utterly.

As long as it lives, hope keeps us alive.

Reason grasps only supposition.

God isn't in that world of shells and colored stones.

God lies beyond that tiny world of reason.

God lives in a world of wonder and miracles.

God lives in a world where imagination takes us.

A world of endless answers.

A world men were born to inhabit, eternally.

A world where demons also dwell.

As I neared the cave that was my destination a dark form fell back, into the larger darkness.

Lately it had begun to stalk me, to whisper.

It offered me disillusionment.

It offered me a sense of futility.

It offered me power. I could become a churchman it told me, the high churchman of nihilism.

Only among men could the lack of any belief become a creed.

Those who know arithmetic may argue whether zero should be part of numbers. About nothing there is no argument.

Nothing is part of the all.

So is infinity.

So too is death and immortality.

I would rather spend eternity in quest of the light than join with the darkness in exchange for power. The demon offers only the mindless adulation of the mindless.

It was only my third visit to the cave of the blind recluse but he recognized my step.

"Good evening, Dhan", he said, as I entered.

I returned his greeting and lit the lamp.

In the cave spoken words seemed to ring more solidly.

In the open air they seemed to dissipate more quickly.

In town they seemed overwhelmed, drowned in other sounds.

I found comfort in the cave.

The recluse was one for whom the almighty was not simply supposition. His own great empathy lent existence to the divine.

After one has rejected fear, empathy is our only way of truly knowing.

I think, therefore I am not.

Thought drove me to doubt.

Doubt drove me from the temple.

I came to Giza bereft, without a reality of my own, until I learned to doubt the things the demon whispered.

In the silence beyond the cry of the legendary bird lies the land of imagination. A place where reason may not enter.

If we peer into things deeply enough, if we ask why often enough, we had best be prepared, one day, to awaken there.

At Giza I learned of the possibilities within impossibilities.

At Giza I learned men create.

Only when a man is free can he create.

To be free is to be free from dogma.

The most enslaving dogma of all is reason.

A man must avoid becoming a collector of shells and colored stones, a purveyor of theories.

Reason was created by men.

Reason cannot reveal anything profound. Reason is a refuge for those who would avoid the profound.

Men of reason cannot take the world very far. Reason cannot create those things the world needs.

There is no magic in reason.

Men of reason go on voyages of discovery.

Men of imagination take the same voyages, without ever leaving home.

"A high pyramid will soon begin its rise from the stone of the structure", I told the recluse, as I found places for the waterbag and food I had brought." It will be a work of art, unless defiled by Khufus' carcass."

"It won't be", he answered quietly. "Arrangements are underway for Khufus' burial elsewhere. His heir is chosen."

"Neither he nor Khufu have any further interest in the structure. It will be completed with little attention from them."

The ability we all have, to see what is yet to come, was far greater in the inhabitant of the cave. He was a seer.

What he said was true.

Internal work was being done in a sloppy manner, some not being done at all. No one seemed to care.

"Were you followed?" he asked me.

"Yes." I answered.

"Once they were servants of the most high", he said. "Now they vie with heaven for power."

"They are more blind than I. There can be no such contest with heaven. The almighty is a blacklander."

The almighty chose love, not power.

He transforms, uplifts.

Demons mutilate, debase.

"It is not yet official", I said, "but the pyramid is to be named, 'Khufu is the horizon'."

"A fitting title." He said.

I understood, and agreed.

If you walk toward the horizon it retreats from you, it also follows behind you, keeping the same distance all around you.

It seemed somewhere, yet was not.

Many will believe they know where Khufu is buried.
Yes, it was a fitting title to give to our pyramid.
Silence flowed into the cave.
We sat and listened while it filled every corner.
Time passed.
The first transformation was complete.

The height of each of the great mastabas had been halved. New mastabas, the new height, had been added.

The inner pyramid rose higher. The outer pyramid, its quarry, rose with it but lost in width.

Abi and Amus no longer worked with us on the structure. They had married Pooths' daughters.

Amus managed the business.

Abi managed land that belonged to the business.

Setis' father also left the job. He lived in town.

He could have lived a life of leisure but chose, instead, to do odd jobs that Amus needed done. He became another asset of the business.

It was Pooth that enjoyed a life of leisure. He and Amus had reversed their roles. Amus now regaled Pooth with stories of his commercial machinations.

Over the years pyramid labor and Pooth had made each member of the team a man of means. As Pooths' wealth grew, so did our own.

On my next visit to the cave, as we sat and listened to the silence, the recluse suddenly broke it, saying:

"A day will come when the creator of heaven and earth will dwell among us."

Silence returned.

It became a stillness. I could have heard the scuttle of a scorpion.

"He will take human form", he resumed, "and dwell among those who most need his love."

"He will dwell among the poor and neglected, among the troubled and despised."

I knew he spoke the truth. As unbelievable as it sounded, I knew it was the truth.

As a miser knows the truth of gold.

As some know only the truth of themselves.

Struggling over the words he next said, "He will wash his disciples feet."

I waited while he regained his composure.

"There is a woman and a man", he said. "Both are disciples."

"The woman does not misunderstand. She knows who the master is. She knows he is divine."

"The man badly misunderstands. He is jealous of the woman. He criticizes her to the master and is admonished."

"He decides to betray the master to his enemies."

He related the story as though he had witnessed it.

I sat and listened and believed.

As some believe that truth lies in reason.

"He will be executed", he told me. "He will be sacrificed for all of us."

He was done. He had shared a nightmare he had seen.

Later, while I walked, I wondered why so much thought of god in our desert land.

Before we became men we possessed the truth.

Before we became men we possessed instincts.

Then we could not hear the demons' voice.

When we became men we heard that voice.

When we gained a soul we became prey.

Men lost their instincts then found imagination.

We then realized what could be.

Men lost their instincts then found doubt.

We then realized what could not be.

Once we had instincts.

Once we had blind belief.

Now we have faith.

Time passed.

A second transformation was complete.

The structure grew ever higher. The mastabas grew ever lower.

Nin left Giza.

He took a good deal of wealth with him. He carried it in the form of receipts for grain. He chose the former nome of Suri in which to settle.

Suri left with him. He carried his own bundle of receipts.

It was sad, saying good-bye to my old friend.

On my next visit to the cave, after I put away the waterbag and provisions. I asked the recluse. "You say the almighty will come to earth, to be mocked and spat upon, to be beaten and cruelly murdered. To what purpose?"

"To redeem us", he said. "To redeem his name."

"Never again will any be able to sacrifice the life of another to him."

"Men will still destroy each other", I said.

Autocrats will command it.

Plutocrats will demand it.

Theocrats will have it.

"None", he answered, "can say it is in his name. He will have redeemed us from that lie."

"How do we obtain redemption by abusing and murdering the divine?" I asked.

"We but murder the man he will be", he said.

"Why be shocked?" he asked.

"His love for us exists. He suffers. He suffers along with all the victims of evil."

"That suffering is our other sin."

Time passed.

A third transformation was complete.

There was a good deal more empty space around the structure. A stoneyard had vanished into it.

The twins left the job. They returned to the river. They took charge of the first ship belonging to the business.

Seti and I remained on the job. We managed to supply all the honest labor the business needed to keep its warehouses, and the warehouses of others, filled.

"In a cult he himself founded," I asked the recluse, "why will there still be so much evil?"

"Love of god is not a cult", he told me.

"Men will still hear the demons' voice."

"Once they were angels. They served the almighty. They served of their own free will. They were created free to choose."

"Then came a time they chose not to serve. They became demons."

"The almighty did not change them into demons. Their own behavior did this. Each thing followed in its course."

"They mocked those angels that yet served and were expelled from heaven by them."

"A desire to frustrate heaven grew within them."

"A great deal can be said for the excitement of living dangerously", the recluse said. "A great deal can be said for excitement."

"You believe that once we were as other creatures, then rose above them."

"We acquired a soul."

"We acquired the ability to sin and feelings of shame and guilt."

"Shame and guilt were a salvation but we had also acquired freedom."

"The ability to free ones self from shame and guilt now existed."

"The ability to make sin a way of life now existed."

"Evil now existed."

Khufu died. The good son was installed as pharaoh. He would build no pyramid at Giza.

The power of the sun cult was further eroded.

Time passed.

The final transformation was complete.

The final ramps are narrow.

The topmost ramps go only to the top of the next-to-the top tier of stone. There is no need for them to go any higher.

In this way the final roadways of that final, highest, mastaba will be lengthened.

They will be longer than the roadways of the mastaba below them.

Each mastaba was now three cubits high but each did not contain the same amount of tiers.

I stood atop the ninety-second one, the highest one.

I looked down on all the other construction.

I looked down on a temple that wore brick designs instead of granite ashlars.

I still could not believe the narrowness of the final ramps.

"This is what became of all that width", I said to Seti, as I swung my arms wide, "we traded it for this, for height."

Seti laughed.

"You'll never change", he said. "When did you first notice, this morning?"

"The magic of creation will always amaze me", I answered.

The height, of any of the inner pyramids' final tiers, cannot be greater than the width of its final ramps and roadways.

These final ramps and roadways are the remaining width of the great step pyramid.

They enclose the stone of the inner pyramid.

Next comes buffer stone.

Buffer stone adds to the girth of the inner pyramid. It creates enough space to emplace the capstone.

The space to accommodate the buffer stone was borrowed from the cleared area around the structure. It will be repaid when the third phase of construction is complete.

Next comes excess stone. This stone will be cut away as a pyramid is formed.

Then comes the area beneath the island of creation.

This area is more than twice as wide as a width of excess stone. That is to say that the capstone will be higher than a final mastaba.

Longer and wider mastabas make for a greater amount of rubble.

That will be no problem.

Rubble will be taken down while other men, more skilled, sculpt a pyramid, one mastaba at a time.

A good deal of stone, with which other structures will be made, will also be taken down the final ramps.

The capstone stood atop the highest mastaba of the structure.

Tomorrow, at sunrise, men will move it a final time. They will move it into position.

Then they will cut away the island of creation, at its base.

There will be a great crowd below, watching that centering. It will mark the end of the second phase of construction.

It will also mark the beginning of three days of celebration before the third and final phase begins.

It was on the second day of that celebration that I learned Anan had vanished. All of his belongings had vanished with him. It was said he had disappeared the night before the centering of the capstone.

I long suspected Anan had savings with Pooth.

I spoke to Amus.

Pooth and Amus each had an irrepressible sense of humor but as men of business both were very serious.

"What I now tell you, I tell you in confidence", Amus said to me.

For many years Anan had sizeable savings with Pooth. It was the value of his fathers' land.

His father had sold that land and followed his son to Giza. He had deposited the proceeds from that sale with Pooth, in his sons' name.

It was Anans' worth that had given us our start in business.

From the beginning Anan had been a partner in our enterprise. None had known, except Pooth and Amus.

"By arrangement", Amus said, 'he came the night before the centering of the capstone. He had other men with him. I gave him receipts. They were all for gold."

"The last thing he said was that I was to ask you to look after his simpleminded father. He said you owed him that much, for making you a man of means."

Amus then asked me, "The hermit you visit, the one some say is holy, is he Anans' father?"

"Yes", I answered. That the recluse was Anans' father came as no surprise to me. I had guessed as much.

Once more, through the magic in papyrus, in the land of papyrus, we move on.

After an exchange of greetings, as I lit the lamp, I asked the recluse. "Did you know your son left Giza?"

"I thought he might", he told me.

I occupied myself putting away those things I had brought and gathering together those I would carry away.

"Khufus' heir will not live long", he told me.

"Will it be murder?" I asked.

"Yes', he answered. "Few will know. Only the guilty."

"Another son will rule. A madman dedicated to his own madness, devoted to his assumed divinity."

"He will try building an even greater pyramid at Giza. He will fail."

"When he dies his son will have the stone of his incomplete, high step pyramid and high pyramid returned for other use. His middle step pyramid will be transformed."

Silence ensued.

"A time of great evil is coming", the recluse said. "A time of human sacrifice."

"Many will go to their death willingly, believing it to be heavens' wish. Evil churchmen will corrupt them."

"I told my son this evil was coming. I believe this is why he left Giza."

'He hopes to sow seeds and reap a harvest of destruction."

"One day that destruction will come. Tombs will be opened, corpses torn apart and strewn about the land."

A great many men will pick up rocks. They will destroy the sun cult. They will raze its unholy city. They will destroy its most precious relics. They will pound them into dust."

"No matter how hard the stone from which they are made, they will be pounded into dust."

It was quiet for some time.

I felt a sadness. With the help of heaven we had defeated the sun cult. Our victory would be short lived. Our nation would be plunged into darkness.

Would the ritual cannibalism, that is so often a part of human sacrifice in other nations, also come to pass I wondered?

Do redlanders satisfy their craving for power, for a time, by eating a victims' flesh, by drinking a victims' blood?

"Men who love freedom will always be the enemy of those who crave power", the recluse said, "it is their ka." I knew he meant his son.

Somewhere is a peaceful island.

Here men of reason come ashore to display their shells and colored stones, as male birds display their bright feathers.

The island is a refuge.

A refuge from vicissitudes and contradictions.

Hope is part of existence.

The recluse told me the divine will come to earth in answer to hope, in answer to prayer.

Reason will have no savior.

Faith is part of existence.

As much as reason is an attack on faith, just so much will it be too bad for reason.

When the almighty gave us doubt he left us unable to believe in anything, seeking to comprehend everything.

Before we can truly know anything, we must comprehend everything.

At times, awash in despair, a human voice cries out.

I'll give you back the eternal hunger.

I'll give you back the endless search.

Give me back the oblivion of belief.

Give me back an unending present.

The Ninth Scroll

In the beginning was a plan and a wilderness of stone.

A wide road will come up from the river and run into the wilderness of stone.

It will be called the road of truth.

It will bisect the north face of each of the coming pyramids.

It will grow wider with each structure.

The rough granite beams of the relieving chambers and the smooth granite roofing beams of the high chamber will travel from the river, to the structure, along this road.

One day a roofed concourse, for a funeral procession, will be constructed down its length.

Four narrow roads will also be constructed.

Each will be named for a cardinal direction.

Each be alined with one corner of the coming structures.

Four straight paths will connect these four corner roads.

Each path will be widened into a cleared area. A cleared area that will surround a structure.

As leveling crews widen the paths others will search for four geometric points, one at each corner of a structure.

None will ever be found.

It would take more than one lifetime to locate one.

That legendary line, outside such a point, can only one other line, less legendary, be drawn through that point and parallel to the legendary one?

Can an infinite number of lines, parallel to that legendary line, be drawn through that point?

It is a puzzle.

Perhaps eternal.

Walls of water will be constructed.

One of the two solid walls, that hold such a watery one, will be coated with white mortar.

The wall of water will be dyed.

One solid wall will be removed.

The line remaining, on the whitened wall, will tell workers, with jeweled drills, just how far down to drill.

While leveling is being done there will be a good deal of work for masons and their laborers.

Masons build up low areas.

They absorb outcroppings into the structure.

They construct rubbleboxes.

A rubblebox is constructed of rubble and mortar. As it is being made it will be filled with rubble.

The west ramps will be the down ramps for rubble from the structure and for the night crew that grease beneath the wooden rails with tallow and remove jars of night soil and other waste.

The south ramps will be their up ramps.

Rubble comes up from the quarry in rubbleboxes that go from the crest of the east road, past the south ramp, then on to the dump.

Men will use a branch of the west road to lever rubbleboxes to the cliff above the dump, then off of that cliff.

Pyramid stone will be sawn. It will be sawn with jeweled saws.

Cuts will be sawn into corestone. Wedges can then be used to crack that corestone into tierstone.

Only with saws can a pyramid be constructed.

Where it meets the pyramid the east road becomes a receiving platform.

Here stands the dispatcher.

He sees to it that the working tier is supplied with the stone that it needs.

He is a man most swear is half crazy, others say he is more so.

When things go well he roars. When they do not his voice gets louder.

Ten hours a day he bellows.

The third and final phase of construction will begin with men, atop the highest mastaba, quarrying along the remaining ramps and roadways to the end.

They will quarry onto the next face.

They will remove buffer stone and excess stone on that face.

The remaining casing stone will be cut into steps, then cut into smaller steps.

While this is being done stone will be stripped from the next, lower, mastaba.

Limestone is mutable; the structure, maleable.

A pyramid is not simply constructed.

It is sculpted.

The great pyramid was complete.

So too was its pedestal.

A pedestal was not constructed until a pyramid was complete.

The great pyramid was surrounded by a white wall.

The ground between the white wall and the immense white structure was paved with white stone.

At last I am done with the pyramid, but not with life. The scribbling continues.

Creation is never finished.

We each take part in the creation of our nation, as the evil angels took part in the creation of their own.

Honor, won by our nation in the world, was lost in the magical pictures on our walls. Too many began to strut.

Honor became vainglory.

The good pharaoh was not depicted pulling the beards of our enemies.

The good pharaoh was not popular.

The good pharaoh would build no pyramid. He planned a mastaba. He would be buried beneath it, as mortal pharaohs were once buried.

The good pharaoh would not build at Giza. He would build to the north, on a height overlooking Giza.

All this would be as cold water thrown on the fiery preaching of the sun cult.

Churchmen of the sun cult behaved as though what the good pharaoh believed mattered not.

Every pharaoh is divine, they continued to preach.

People listened and worshipped other men.

Why does our nation hold sway over so many other nations they asked, in lower tones?

People listened and strut.

Too many now accept that the easiest road is the one that leads to truth. They call this pragmatism.

Their truth lies not very far away.

It lies only as far as they can see.

Seti moved north with the new construction.

He followed his god. His god is justice.

I told no one the recluse had told me the life of the good pharaoh would be a short one.

Hemiunu and I had a final meeting. It took place in the park where we first met. We sat on the same stone.

Before the conversation began he had me promise that I would not breathe a word of what he told me to a living soul.

I so promised. I intend to keep that promise.

Only you, not yet born, will be told what he said.

Khufu was buried in the south of our nation.

He was buried, secretly, in the city holy to the god of good and evil, as Hemiunu called the divine.

Hemiunu was one of the burial party.

By Khufus' own wish, large stone statues of him remained at Giza. After a man's death such statues often serve as a resting place for his soul.

Only a small ivory statue of him, seated on a throne, was taken south. It showed him wearing the crown of ruler of the south.

It was placed in a niche in a temple holy to the god of righteousness, Hemiunu's god of good and evil.

Let them come to Giza and leave gifts for Khufu.

Let them come and leave their bribes.

Let them wander among granite statues, empty of him, at Giza.

Those closest to Khufu in life know, if one would speak to Khufu in death, one must journey to another place, to a different temple.

A few drops of the water of chaos, sprinkled on the great pyramid, wrought great changes in that structure.

Those drops also wrought great changes in Khufu.

This Hemiunu thought I should know. This, I feel, you should also know.

There are other things, I feel, I should now point out to you.

The low chamber was excavated in a three hundred and sixty degree manner, this being the most efficient way when time is a concern.

The place where a sarcophagus, made of bedrock, would have rested beneath the pyramids' apex was never reached or excavated.

As was intended.

Nor was the middle chamber ever cleared of all fillstone.

The place where an unfinished sarcophagus, made of limestone, rests beneath the pyramids' apex is yet filled with stone.

As was intended.

Only in the remaining chamber, the high chamber, could a resting place now be found beneath the sacred stone. But fate, or something else took a hand.

The high chamber was moved. It was moved far off center.

Now none could find a resting place beneath the sacred stone.

There is something else that should have been made more clear to you.

A step pyramids' ramps are not actual stone slopes.

They are a slope of small stone steps.

There is a reason for this.

When constructing a larger pyramid, around and over an existing step pyramid, old ramps will not be disturbed. They will be closed.

They will be closed by using stone with an upside-down slope of steps on one face. Stone that will mesh with the closed ramp.

Such stone will be employed along each ramp, tier by tier, after the wooden rails are removed.

A structures inclines will be closed with fillstone.

All this is done so that if construction of the larger pyramid is aborted, and all of its stone removed, the smaller structure can then be continued to completion.

The recluse died. I saw to his burial.

I accompanied his body south. He was buried in the same city as Khufu, the city holy to the god of righteousness.

I still go to the cave. It is a refuge.

It is a place where one may sit quietly and look into ones' deepest being.

It is a place to which many would have to be dragged, kicking and screaming.

I wander in a dream.

I roam inside a room filled with ancient stone tablets.

Some are titled RIGHT.

Some are titled WRONG.

Time has eroded what was once written beneath these words.

I roam inside myself. I roam in a room once called instinct.

We are no longer part of a pack.

We each now have freedom.

Freedom can sometimes take one to a place whose inhabitants reject all boundries, especially moral ones.

A great deal can be said for the excitement of living dangerously.

A great deal can be said for excitement.

Without righteousness life would be mindless chaos.

Foreigners come to Giza.

They come to marvel at a pyramid.

One is overwhelmed by the immense, white simplicity.

It is not a tomb. We saw to that.

It is a faith, in stone.

What towering monuments, to their own faith, will future ages build?

Foreigners come to Giza. I married one of them.

Her name is Helen. She is from the greek speaking lands. We have a son.

It is strange I married Helen. A woman whose people are such worshippers of reason, such believers in cause.

In any storm who knows how many grains of sand the great green washes ashore.

Who knows how many grains the great waves suck back into its maw?

Who cares?

Truth is not about cause. Truth is about creation.

Creation is inspiration, miracles.

Creation is music, magic.

Creation was not caused.

Creation can have no cause.

Cause was not yet created.

Creation needs but a creator.

We had best be careful. The almighty may destroy our world and create another.

A world based on a new arithmetic. A world based on a logic we cannot understand, but new creatures do.

He may already have done so.

Helens' god created cause. Our god created Helens'.

Our god, uncaused, is the creator of cause and lord over chaos.

He is god over the all.

The almighty is not there to boil our water.

The waters of chaos yet exist. They exist in randomness and chance.

Our island exists within that sea of probabilities and improbabilities.

Greek philosophers believe that what is true must also be logical.

They also believe that because a thing is logical it is true.

Yet among them dwell those who seek escape from the rational. They practice rights, participate in ceremonies, designed only to excite.

Philosophy, like solitary sex, excites one man.

Another prefers depravity.

Some even put their faith in doubt, they thrill in creating it. They thrill in undermining beliefs, the beliefs of others.

Evil has its own logic.

Excitation is the consequence of belief.

Being craves excitation.

At carnival time some seek the giant slides.

Some seek the giant swings.

Some seek sex.

Some antelope even tempt the lion to give chase.

Being craves excitation.

Which came first, physical existence or the desire for it?

Which came first, a need to sing or the human voice?

What created being?

Not eyes or any other part of us; what created being, no more substantial than music?

What created being, if not that greater being?

I believe we underwent changes.

I believe we rose above other creatures.

I believe we acquired freedom.

Only those who know freedom can doubt, can wonder.

Only those who know freedom can imagine, can speculate.

Only those who do both can create.

Now we know transcendental aloneness.

Our faith is an answer to such aloneness.

Our faith is, perhaps, the answer to gods'.

Suppose there was a time we did not yet have eyes? How do blind creatures come to see?

What arranged for the delivery of scenes to being?

Was it the invisible self?

The self that wakes one because it has heard a strange sound.

The self that wakes one because it has become too quiet.

Many now view themselves as possessing another, unconscious, mind. Many no longer view themselves as possessing another, greater, self.

To fly, did birds reach down inside themselves to create a feather?

Did they, and hope, give birth to it?

Is hope the most primitive form of prayer?

Is faith our feather?

Some, not all, believe others exist.

Some, not all, believe a dragon once existed.

Some, not all, believe a greater existence exists.

How did we obtain a soul? It is the soul that seeks an answer.

The good pharaoh died.

Work ceased on his mastaba.

Seti returned to Giza.

The evil son reigned. He planned to build a pyramid at Giza. He planned to make it greater than his fathers'.

It is said it will be a true pyramid, not a tomb. None will be buried within it.

Men set to work, constructing a road of truth.

Far to the north, in delta country, men immobilized the guards around a granary one night. By dawn, all the grain was gone.

Hunger in that region was assuaged.

In a southern nome a ship, loading corn, was hijacked. Days later it was found, aground and emptied of cargo.

Often, when I learned of such an incident, I thought of Anan and his men.

Pooth died.

Setis' father followed.

"I'm retiring from construction", Seti told me, soon after his fathers' burial.

I knew why. I waited for him to tell me.

"I spoke to Amus", he said. "I will take charge of the warehouses. With the new construction at Giza there will be more activity in them."

He said no more.

The wealth of the royal family was great, so was their squandering of it. A large part of what remained was now to be lavished on a new pyramid.

Those whose task it was sought ways to economize.

Cheap labor from the puppet states was one way.

The foreigners would be paid a lot less.

Our own workers would be given a little more.

The scheme was presented to our workers.

They agreed to it.

Those workers who presented the scheme to the other workers, for their approval, became the new leaders.

Where they would lead all but fools knew.

Many of those in our village and towns, without work, will suffer. There will be less work for such as them.

So Seti took charge of the warehouses.

Anan returned. He looked careworn.

He had lost much of the hair from atop his head.

He seemed to have aged more than the number of years that had past since last we met.

I learned he had traveled the land speaking with church-men of the sun cult. He had traveled in the guise of a be-liever in their plans. He did this at Hemiunus' request.

So, he had struck no blow against the overt greed of those corrupting our nation.

"I lost men", he told me. "When they found I had influence with the sun cult they asked I use that influence in their behalf. They wished to join the elite forces. They were of no consequence", he told me.

"Have you yet seen the statues of gods with animal heads", he asked.

I told him that I had.

"More and more the churchmen of the sun cult preach the new doctrine. They preach that men are no more than beasts", he told me.

That is a lie.

Beasts have sight.

Men have vision.

Churchmen of the sun cult will prey on that vision.

They will prey on peoples' fear of that all devouring nature that lurks beneath the claw and talon. They will seek to turn fear into worship, into reverence for beastial power.

Beastial gods of lust and appetite, beastial gods to promote beastial behavior.

Before they can sacrifice men to obscene gods, as some other nations do, they must first debase them.

There are those who smell no flowers.

Who hear no poetry.

Who feel no music.

Who see no creation.

Who know no wonders.

They look to chance. Their god is a god of accidents.

All men have a god. It is our nature.

There is a blind fish. A fish that went to live in the darkness of a cave.

It lost the use of its eyes.

One day, in darkness, that fish will again see. Just as the cave bat has found another way to see.

It is hope that creates these miracles.

Without hope things remain blind.

Nor can hope achieve anything by itself, alone.

Another must be there to respond, to answer that prayer.

Anan and I arranged to meet again.

Anan still retained an interest in the business.

It was time he knew the state of things. It was time he was told what the group, once a team, had now become.

There were many groups in our nation determined to bring about the eclipse of the sun cult.

Our group, the Giza group, was but one.

Nin and Seti, in their nome, were part of such a group.

The twins, in their travels up and down the river, were couriers between groups.

The leader of each group needed patience. The struggle would be a long one.

It would be won by our children or by their children. The leader of each group needed to know this.

It was why I was chosen to lead the Giza group.

What can pharaoh offer us?

Faith that we create better weapons than our ever increasing enemies?

What will happen when an enemies' weapon becomes truth?

Truth would sweep all before it, even the armies of pharaoh.

So could inspiration.

Far to the south primitive men abide.

Often they begin a hunt by capturing a pair of venomous snakes. They choose a breed that's cranky.

They punch holes in the tails of both snakes. They then tie the two together.

They place the pair in the tall grass, by a waterhole. A waterhole where game comes to drink.

Then they sit and wait.

Man is a most dangerous predator. Even the lion does not realize it is being hunted until it falls into the pit.

Thought is a function, only a function, of being. It is like digestion, only conscious.

Thinking is hunger, a hunt.

Knowing is satiaty, a stupor.

Being is behind that thinking.

Being, using thought, will always seek to find something to justify its faith.

Faith is our feather, our fin, our fang and claw.

It has given us immortality.

We need hope.

For it is doubt that arouses ones' faith and only doubt that can destroy it.

We need charity.

For the struggle will not be against those who have lost faith in god, or against those who never had any.

The struggle will be for them.

The recluse once told me a man will come. A man who will preach just such a doctrine. A man who will also be our creator.

Only a god who would kneel and wash his disciples' feet is worthy of our veneration.

Only such a one should judge us. To such a one all, each of us, are redlanders.

Had he chosen the aloneness of faith in self, for himself, as Anan had done, there would be no miracles.

That miracle called hope, a miracle he himself created, would not exist.

An insect depends on numbers for its survival, meaning the survival of its kind.

Man depends on uniqueness for his survival, meaning the survival of his kind.

Man has adapted a soul as a means for that survival.

Perhaps our ancient ka became that soul.

Perhaps, when we became free, freedom transformed it.

Perhaps heaven did.

Perhaps, there is only perhaps.

Perhaps, for creators, there can only be perhaps.

I wander through a vast library.

I wander among scrolls that prove it is so and scrolls that show it is not.

A whore called reason is the librarian.

She sleeps with every man.

She tells each what they wish to hear.

There's no escaping her.

Any explainer who would submit a scroll to her endless collection must do so to her.

Explaining is a fraud explainers perpetrate on those seeking answers.

Magicians create.

Magicians create what, once, none knew. They turn the little we know into a good deal more.

From whence comes this ability to create?

There is no rational explanation for the miracle of creation. Thinking has limits.

Ages come and go.

There will come a time when reason becomes a cult. A time when knowledge is called power.

An age without wisdom. An age of shells and colored stones.

A time when men will build no pyramids.

A time when theories become dogma, heretics are excommunicated, nonbelievers mocked.

We near the end of our journey.

I have waited, until now, to tell you a story.

While I was at school I had a friend whose study was astrology. One night, without telling anyone, he packed his things and left school. When I had an opportunity I spoke to his teacher.

His teacher told me that my friend had come to believe the earth and planets moved about the sun, in perfect circles.

It was not uncommon for a bright student to fall into this trap, his teacher told me.

If one made a study of the heavenly charts, with this theory in mind, it would almost seem true. Almost, the teacher said, but not quite.

If one attempted charts as accurate as the true charts, using this theory, they would fail.

The charts would not be able to predict even an eclipse. They would be worthless.

My friend, the teacher said, found it impossible to climb out of the trap he had fallen into. His studies suffered.

Often he argued, loudly, that crystal rails, on which the earth and planets rode, circled the sun. The rails, being crystal, were invisible.

He had been going mad, the teacher told me.

Embarrassed, I did not tell the teacher that my friend had told me about the crystal rails. I knew little about astrology. I had believed him.

One day I will hide these scrolls in the cave where the recluse once dwelt.

I will put them inside a clay jar and bury it.

Before his death the recluse told me it would be thousands of years before they were found.

By then, I am certain, that man will have come. The man who will make those invisible rails visible.

He will show they exist.

By then, perhaps, that other man will also have come. The one who will show such rails cannot be.

Imagination can provide more than one answer.

It just provides them in different ages.

There is power in things but the magic lies in us. The journey you have taken has been a journey into that magic.

Do not seek truth in reason.

Truth can only be found in faith, in hope and in empathy.

It can only be found in the substances of human existence.

For you the journey is ended.

For us another journey is underway.

We have a new reality to create.

Afterword

A later scroll. Written by a descendant of Dhan.

In our land comes a holy time.

A time when sanctified statues are exchanged between two cities.

A statue of the sun god leaves its city in the north and journeys south.

A statue of the god of righteousness leaves its city in the south and journeys north.

Each will be welcomed into the opposite city. As followers of that god are welcomed.

It was a time for all to realize that there is but one god, with many names.

It became a time for carnival.

It became a time for drunkenness.

It became a time for excitation.

It became a time when those who built the pyramids debased themselves, while others looked on in amusement.

What came to pass none foresaw.

Not even the leaders of the secret resistance foresaw what came to pass.

To north and to south ships and barges of polished wood, gilded with gold and trimmed in ivory, gathered.

To the south a bejeweled and clothed, sanctified, statue was placed aboard a ceremonial barge.

To the north the same was done for a bejeweled and clothed idol.

I can testify.

At that time none knew the idol was journeying south to be cleansed, to be resanctified.

There comes a time when heaven takes matters into its own hands.

There comes a time when god himself moves mens' souls.

Men sicken of being instruments of injustice.

Men sicken of being tiles in other mens' games.

Men sicken of mindless cruelties that some endlessly enjoy.

Unless they have come to see the most beastial cruelties as manliness, men sicken.

When men sicken of tyranny what can tyrants do?

To north and to south, towns and villages in advance of each flotilla prepared for carnival.

To the south the growing numbers that advanced with the flotilla were too serious for carnival.

As their number grew they became somber.

Barrels of wine and beer, put out for carnival, were broken open, their contents spilled upon the ground.

To the north, as more and more men became conscious of the purpose of the journey south, the idol was cloaked from sight.

Churchmen that accompanied the idol were replaced.

Priests of the old sun cult joined the idol on its journey south. They would see to its purification.

To the south, more and more men merged with the crowd.

The hundreds became thousands, the thousands became tens of thousands.

All moved north.

Early on, the towns ahead ceased to prepare for carnival. Those who lived for carnival melted away.

Towns now prepared for an army.

It was an army without banners. It bore the sanctified statue. It bore the statue of the god of righteousness.

It was an army without fear.

It was an army of believers.

Among its many thousands were veterans. At first they had carried their swords concealed. Now they wore them at their sides.

Among its many thousands were hunters. Many carried bows. Many carried spears. Others carried slings.

When the time came most would rely on themselves, on each other, and on god.

The army sorted itself out. Leaders were chosen.

Before they arrived in the city of the sun they would traverse the capital.

The two lay close together.

They would lay waste to both.

There comes a time when men must destroy.

To preserve righteousness what is unrighteous must be destroyed.

Many believe the struggle to evict the evil angels from heaven ended in victory, long ago.

Many believe that struggle continues and will continue until the end of the world.

There is the physical journey.

There is the grave.

There is a spiritual journey.

There is a destination beyond that grave.

If imagination could sort through those infinite answers that heaven has provided, it would find no answers to our deepest questions but faith.

Were there any other answer heaven would have provided it.

There is no other answer.

Truth lies in the magic of creation, not in its arithmetic.

God lies in the magic of faith, and hope, and love, not in cause and effect.

We go where faith takes us.

Faith is the philosophers' stone.

www.ingramcontent.com/pod-product-compliance
Lightning Source LLC
Chambersburg PA
CBHW051319170526
45166CB00002B/602